SO-AXS-301

THINK SMALL. *WIN BIG.*

INVESTING
IN
NANOTECHNOLOGY

Profiles Over 100 Leading
Nanotechnology Companies

JACK ULDRICH
author of *The Next Big Thing Is Really Small*

PLATINUM PRESS™

Avon, Massachusetts

Copyright ©2006, Jack Uldrich.
All rights reserved. This book, or parts thereof, may not be reproduced in any form
without permission from the publisher; exceptions are made for brief excerpts used in
published reviews.

Published by
Platinum Press™, an imprint of Adams Media,
an F+W Publications Company
57 Littlefield Street, Avon, MA 02322. U.S.A.
www.adamsmedia.com

Platinum Press™ is a trademark of F+W Publications, Inc.

ISBN: 1-59337-408-9

Printed in the United States of America.

J   I   H   G   F   E   D   C   B   A

Library of Congress Cataloging-in-Publication Data
Uldrich, Jack
Investing in nanotechnology : think small, win big / by Jack Uldrich.
p.   cm.
ISBN 1-59337-408-9
1. Stocks. 2. Nanotechnology. I. Title.
HG4661.U43 2006
332.63'22—dc22
2005026066

This publication is designed to provide accurate and authoritative information with regard to the subject matter covered. It is sold with the understanding that the publisher is not engaged in rendering
legal, accounting, or other professional advice. If legal advice or other expert assistance is required,
the services of a competent professional person should be sought.
> —From a *Declaration of Principles* jointly adopted by a
> Committee of the American Bar Association and a
> Committee of Publishers and Associations

Many of the designations used by manufacturers and sellers to distinguish their product are claimed
as trademarks. Where those designations appear in this book and Adams Media was aware of a
trademark claim, the designations have been printed with initial capital letters.

This book is available at quantity discounts for bulk purchases.
For information, please call 1-800-872-5627.

Contents

"The subject of nanotechnology is vast, the potential is immense, and its study is a wonderful endeavor to embark upon . . . but now is the time to do the homework and become informed investors."

—Carl Wherrett & John Yelovich of MotleyFool.com

Preface

If you think that investing in nanotechnology is a quick, easy road to riches, this book is *not* for you. Nanotechnology will create a lot of new wealth, and it will destroy a lot of old wealth. But it will not, as a general rule, do these things overnight. Rather, it will do so over a period of years, and accordingly, investors of nanotechnology will need to demonstrate some patience.

A few historical examples are worth bearing in mind. The first is the semiconductor industry. The first transistor—the basis of today's computer industry—was created in 1947. It was not until the creation of the integrated circuit, eleven years later, that the potential of the industry began to become evident to some very farsighted investors. Those investors did not see a return on their investment until the mid-1970s when integrated circuits began moving out of very specialized applications and were incorporated into some familiar commercial items. But those investors didn't really see a handsome return on their investment until the manufacturing processes grew so efficient and cost-effective that integrated circuits started to drive the growth of entirely new markets in personal computers, cell phones, and a host of other popular consumer products. The time frame between the first transistor and the semiconductor industry's dominance was roughly forty years.

The biotechnology industry followed a similar pattern. Today's biotech giants, Genentech and AMGEN, started nearly two decades ago. But only recently have they become profitable.

These two examples are offered merely to control the "irrational exuberances" of any investors hoping to strike it rich in nanotechnology tomorrow. The more important point is that both industries did become profitable in the long run and the patient investor was rewarded.

This is not to imply, however, that nanotechnology investors will have to wait forty or even twenty years to begin to reap profits from their investments in nanotechnology. This is true because technology is experiencing exponential growth, while at the same time the span between when a paradigm-shifting scientific or technological advance is first discovered and when it finally makes its appearance in commercial products and the marketplace at large is experiencing a corresponding reduction.

The shortening time frame of the profitability of the semiconductor and biotech industries—from forty years to twenty years—may appear to be just a coincidence; it is not. Ray Kurzweil in his book, *The Age of the Spiritual Machine,* provides a series of examples showing that the time between the development of a new technology and its widespread acceptance by society (defined as 25 percent of the population using the device) has been consistently shrinking. For instance, from the time the phone was first created by Alexander Graham Bell in 1876, it took thirty-five years before one-quarter of the homes in the United States had one. The television took only twenty-five years. The computer took sixteen years, the cell phone took twelve years, and the Internet only took seven years.

I mention these facts because nanotechnology is going to enable a host of new materials, medical devices, energy-related devices, and drugs (which will likely be delayed due to FDA approval). Many of these products are going to be on the market sooner rather than later—many by the end of the decade.

Human tendency is to assume linearity. That is, most people assume progress will proceed in a prescribed, organized, and straightforward fashion. This line of thinking is best exposed with a short quiz. Consider a pond. If a pond lily doubles every day and it takes thirty days to completely cover a pond, on what day will the pond be one-quarter covered? On what day will it be half covered?

Many people respond that the pond will be one-quarter covered in one week and half covered on Day 15. They are wrong because they are guilty of linear thinking. The fact is that by the end of the third week—a full week after many people guess the pond will be half covered—lilies only cover 1/512th of the pond. It is only on Day 28 that the pond becomes one-quarter covered. Of course, it is then half covered the next day, and because it is doubling every day, it is fully covered by Day 30. So why do I tell this story? It is because many people demonstrate the same tendency with respect to the emergence of new technologies that are growing almost exponentially. That is, they *overestimate* the potential for the technology in the short run (i.e., they think the technology will achieve market dominance much sooner than it actually does). When that doesn't happen, they tend to become disappointed, discouraged, or disillusioned. Ironically, it is precisely these tendencies that give rise to the second fallacy; that is, they *underestimate* the long-term potential of the new technology. By the time they finally do grasp how fast it is progressing—say on Day 28 of the earlier example—and hope to capitalize in on the explosive growth, it is too late.

Therefore, I encourage investors to think of the emerging field of nanotechnology as being around "Day 15" in the year 2006. The field has been touted in the mainstream media since 2000, and a number of people have already become disillusioned at its prospects and remain so today because of the relative scarcity of commercial products. The underlying science of nanotechnology, however, is rapidly growing, as is the number of products it is enabling and creating. In order to reap the maximum benefits of nanotechnology over time, investors need to begin learning about and investing in the field today!

Jack Uldrich
Minneapolis, Minnesota
September 2005

"Nanotechnology has a mortal lock on being tomorrow's gold mine. It will produce trillions of dollars in new wealth over the next century. It's sure to reshape every industry it touches . . . computing, materials, health care and so on."

—Rich Karlgaard, publisher of *Forbes*

Chapter 1

Big Thinkers Think Small

In my first book, *The Next Big Thing Is Really Small: How Nanotechnology Will Change the Future of Your Business*, I described nanotechnology as the "willful manipulation of matter at the atomic and molecular level to create better and entirely new materials, devices and systems." As a way of making this definition more practical, I asked the reader to picture a lump of coal and a diamond ring. Both are made out of carbon atoms, and it was the precise arrangement of those atoms that gave each product its distinct look—and value. I stated that "while the science of nanotechnology cannot yet rearrange the carbon atoms . . . to make diamonds, it is advancing rapidly and will be inundating the business world during the next few years." The book was published in March 2003, and less than six months later, two separate nanotechnology companies had rendered my statement obsolete.

In the September 2003 edition of *Wired,* the cover story was entitled "The New Diamond Age." The article highlighted two U.S. companies—Apollo Diamond and Gemesis—that were using two different techniques to rearrange carbon atoms to make diamonds. In fact, they were manufacturing two-carat diamonds for less than $100—with the potential to go as low as $5.

Their story is a good starting point for a book about investing in nanotechnology because it captures the exciting potential of nanotechnology as well as some of the possible pitfalls.

To begin, the opportunity for both companies is vast. The diamond market is a $7 billion industry, and the fact that both companies can now manufacture in just a few days what it has previously taken Mother Nature nearly 3 billion years to produce could quite possibly transform the diamond industry. According to one diamond expert, the new diamonds "have the potential to bankrupt the industry."

They do, but this is where investors need to pay close attention. Their success is contingent on three factors. First, the companies need to demonstrate that they can produce their diamonds in large quantities; second, they will need to convince consumers that man-made diamonds are a suitable replacement for natural diamonds; and, three, they must make sure no one else develops a better or more efficient manufacturing method.

It is the asking—and answering—of such questions that underscores the importance of investors conducting their own due diligence on potential opportunities in nanotechnology. (This subject will be covered in detail in Chapter 2.) Investors need to understand that Apollo Diamond and Gemesis face a real challenge from both DeBeers, the giant South African diamond conglomerate, and the Diamond High Council, the trade association for the diamond industry. Obviously, both entities recognize the threat to their profits and their industry and are taking measures to protect their monopoly. In fact, DeBeers is now supplying jewelers with expensive equipment to help determine whether a diamond is natural or synthetic. If successful (the technology has not yet been proven reliable), the diamond industry may be able to marginalize the "synthetic" diamonds in much the same way as they have done with cubic zirconium. (The term "synthetic" is in quotation marks because the molecular structure of the competing diamonds is identical to mined diamonds. The only difference is that the two nanotechnology companies manufacture their diamonds overnight, while DeBeers mines a product that has been forming over billions of years.)

The diamond industry has also successfully lobbied the U.S. Federal Trade Commission (FTC) to prevent at least one of the companies (Gemesis) from labeling its product a real diamond, and it is trying to do the same with Apollo's. The diamond industry is also likely to orchestrate an advertising campaign to convince consumers that mined diamonds—because of their nearly timeless age—are more symbolic of a person's lasting commitment to a relationship.

Such regulation and marketing efforts may ultimately be successful and could serve to discourage some investors from the two companies. However, as with most investing, there is an upside for the investor who is willing to bear some risk. To wit, Gemesis, the company the FTC has ruled cannot label its product a real diamond, is considering naming its product a "cultured diamond." The idea is taken directly from the pearl industry where the man-made "cultured pearl" has over the past half-century become more valuable than real pearls.

Apollo, on the other hand, because it can precisely manipulate the atomic composition of materials is creating diamonds that are absolutely flawless. Therefore, while the diamond industry may be able to successfully get its diamonds labeled as "synthetic," if Apollo's are clearer, stronger, more beautiful, flawless, and less expensive, it is quite possible that the consumer won't care whether the product was produced over billions of years on the African continent or in a matter of hours in some strip mall outside of Boston.

The difference between Gemesis and Apollo Diamond's technologies leads, indirectly, to the second danger of investing in nanotechnology: Often there are different ways to "build a better mousetrap." Throughout this book, the reader will note that many different nanotechnology companies are working on different approaches to solve the same problem. In the case of Gemesis and Apollo Diamond, it is too soon to determine which company will be superior, but investors need to have some understanding of the underlying technology because it may have implications for additional markets for a company's product or technology.

In the case of Apollo Diamond, the company uses chemical vapor disposition—a process of tweaking the temperature, pressures, and gas concentrations—to build its diamonds. This leads to the possibility of building large diamond wafers capable of being used in the semiconductor industry—an industry that dwarfs the diamond industry in terms of revenues.

This might seem unimportant until one understands that although silicon has many wonderful properties, including its relatively low cost, it has neither the conductive nor heat resistant properties of diamondoid materials. To date, silicon's properties have more than met the needs of the growing semiconductor industry. However, this could soon change. In the near future, silicon is expected to run into some severe physical barriers. As the number of integrated circuits continues to double every twelve to eighteen months, the circuits are running in much closer proximity to one another and at ever greater temperatures.

If this trend continues, there will soon come a time when the circuits get so hot they will simply liquefy silicon. One potential replacement material will be diamondoid materials. In its natural state, diamond is an inherent insulator, meaning that it doesn't conduct electricity. However, if boron atoms can be injected into the diamond, it can create a positive charge. Researchers are now experimenting with how to give diamondoid materials a negative charge as well. If successful, they will have created a p-n junction—an essential component for making an integrated circuit. The result could create a huge opportunity for Apollo Diamond. (Note: There are other technologies that could supplant this diamondoid material, and I don't mean to imply this scenario is a given. It is merely a possibility.)

The last danger is simply that a new method for developing diamondoid material may be developed. For instance, in May 2005, it was announced that researchers at the Carnegie Institution's Geophysical Laboratory had produced a 10-carat, half-inch thick single-crystal diamond at a growth rate of 100 micrometers per hour—a five-fold improvement over other commercially produced diamonds.

More Than Diamonds

If it is possible to manipulate carbon atoms into diamonds, what else is then possible in the field of material sciences? The answer: quite a bit. From textiles and glass to plastics and steel, nanotechnology is poised to usher in what some experts are calling the "Next Industrial Revolution." Nano-Tex, a subsidiary of Burlington Industries, has been manufacturing stain-resistant pants for three years. More recently, Pilkington, Asahi Glass, and Nanogate have all announced that they are manufacturing self-cleaning glass that utilize nanoparticles. In 2004, GM unveiled its new Chevy Impala, which is made out of super lightweight, scratch and dent resistant nanocomposites; and Nanocor, a subsidiary of Amcol, is manufacturing tons of nanoparticles for everything from more gas impermeable plastic beer bottles to lighter food packaging.

But these developments are the tip of the proverbial iceberg. DuPont is working with the U.S. Army and the Institute for Soldiering Nanotechnologies to develop clothing that is capable of monitoring the health of the individual user. One of its products will be an advanced uniform for U.S. soldiers capable of generating its own power and maybe even camouflaging the soldier to match any given environment. DuPont is investing in the research in the expectation that the advances will be commercially viable. Just imagine the market for clothing that helps power a laptop computer, monitors the user's health, or changes color on demand?

Are such expectations realistic? The answer is yes. In December 2003, President Bush signed into law the $3.7 billion National Nanotechnology Initiative. It was the largest government funded science initiative since President Kennedy authorized the Apollo Space program. The five-year program is designed to ensure the United States remains the world leader in the race to develop and commercialize nanotechnology.

It is a race in which the United States is neither currently ahead, nor predetermined to win. In the past three years, Japan, South Korea, Singapore, Taiwan, Israel, Canada, and the European Union have also established well-funded nanotechnology initiatives. For the first time in recent history, many foreign students are now returning to their native homeland after receiving their masters and PhD degrees from American institutions of higher learning. This is because their home countries are now making it attractive to do so by offering higher salaries and providing the opportunity to work in state-of-the-art nano-technology research centers. This development is important for investors because it reinforces the message that in order to profit from nanotechnology they need an investment horizon that spans the globe.

Within the next five years, it is estimated that worldwide investment in the field of nanotechnology will exceed $10 billion. The scale of this investment represents a host of both problems and opportunities for the individual investor. On the positive side of the ledger, the university and federal government labs that are receiving the bulk of this funding will employ the money not only on basic research, but also on developing cutting-edge technology that promises many exciting commercial opportunities. It is these developments which, in turn, are most likely to receive the venture capital funding necessary to facilitate long-term commercial opportunities. The sheer magnitude of money being invested in the field also offers investors their first and, arguably, safest bet to profit from nanotechnology, and that is by investing in those companies that are supplying the necessary equipment and raw materials to the nascent field. (Chapter 3 will cover the leading equipment suppliers, and Chapter 4 will cover the top nanomaterials companies.)

Why It Is Important

As funding for nanotechnology increases and the term becomes increasingly popular, investors need to understand what it is. While the earlier example of the coal and the diamond may be useful, it is also important to have a deeper understanding of the term. The National Science Foundation states that "nano-technology is research and technology development at the atomic, molecular, or macromolecular levels, in the length scale of approximately 1–100 nanometers, to provide fundamental understanding of phenomena and materials at the nanoscale and to create structures, devices, and systems that have novel properties and functions because of their small size." The sentence is a mouthful, but if the statement is broken down into separate parts, it becomes a little more digestible. The two most important components of nanotechnology are its novel properties and its small size.

Let's begin with the novel properties. Once materials are reduced in size to the neighborhood of 100 nanometers, they begin to demonstrate entirely new properties. For instance, they are stronger, lighter, more conducive, or have enhanced optical or magnetic properties. Again, a few concrete examples may help make this clearer. At the macro level, carbon is horribly uncon-ductive. At the nanoscale, however, carbon nanotubes offer vir-tually no resistance. In fact, they offer so little resistance that one Nobel-winning scientist has speculated that carbon nanotubes could be used to produce "quantum wires"—a wire no more than a centimeter in diameter capable of transmitting over a terawatt of energy. On a more practical level, the property suggests that carbon nanotubes may also be an integral component of next-generation semiconductor devices.

Small materials also have an unusually large surface-to-area ratio. This property means that more of the material is exposed and provides nanoparticles with a decided advantage in the area of creating more effective catalysts. A number of companies in the energy industry, including Halliburton, Engelhard, Exxon-Mobil, and Headwaters, are already exploiting this property to produce better, cleaner, and more profitable oil and gas.

The second characteristic that the National Science Foundation identified with regard to nanotechnology is its small size. The most common analogy offered is that a nanometer is 1/80,000th the width of a human hair. A more accurate definition is that a nanometer is roughly the width of 10 hydrogen atoms strung together. Neither definition is particularly useful to the average investor. But what is important to know is this: Material and devices that are less than 100 nanometers, in addition to having the aforementioned unique properties, are also roughly the same size as DNA and viruses. This suggests that they may be very useful in interacting with—and helping better understand—the human body. After all, the human body operates at the nanoscale, and if doctors, medical device companies, and pharmaceutical companies want to effectively treat the body, they will need to begin diagnosing and preventing disease at the nanoscale.

"N" Is for Nanotechnology: Corporate Investment

In August 2003, Hewlett-Packard began airing a rather remarkable commercial. The commercial began with the simple statement: "N is for nanotechnology." It then went on to briefly explain what nanotechnology is and then launched into an assessment of future developments that will be made possible because of nanotechnology. Some of the items Hewlett-Packard listed were lightbulbs that never burn out, cars that can think, and T-shirts capable of giving directions. Near the end of the commercial, the company even threw in this little kicker: "and cell phones so small an ant could use them."

Undoubtedly, some of you are thinking, "Why would an ant need a cell phone?" It's a legitimate question, but those asking it fall prey to the folly of Harry Warner, the former CEO of Warner Studios, who famously asked in the early stages of sound movies: "Who the hell wants to hear actors talk?" (The answer relates to the possibility that Hewlett-Packard could potentially make cell phones so small they could be embedded directly into clothing or the walls in a home.)

What is amazing about the commercial is that Hewlett-Packard even aired it at all. Corporations, especially large ones, tend to be fairly conservative in how they portray themselves to the public and don't usually go out on a limb about future products unless they are fairly confident they can back up their claims. Yet, the fact that Hewlett-Packard chose to spend a good chunk of money running commercials telling the public that it is working on data storage "devices that can store every book ever written," "cars that can think," and "T-shirts that can give directions" is a testament to either its hubris or the status of its research and development. (Given the company's announcement in early 2005 that it had taken another tangible step toward molecular electronics, I am inclined toward the latter.) Only time will tell which it is, but at a minimum, it should put the public on notice about what the near-term future may hold in store.

Hewlett-Packard, while perhaps the largest public corporation to publicly tout its nanotechnology research and development, is by no means alone. Chapter 5 will explore the exciting work being done at today's largest companies, such as General Electric, 3M, Intel, Hitachi, and BASF.

Hype Versus Hope

The fact that you are reading this book suggests that to some degree you are already aware of some of the promise surrounding the nanotechnology industry. Jeff Bezos, founder of Amazon. com, has said, "If I were just setting out today to make the drive to the West Coast to start a new business, I would be looking at . . . nanotechnology." Steve Jurvetson, a principle partner in Draper Fisher Jurvetson and one of Silicon Valley's most prominent venture capitalists, has said, "Nanotech is the next great technology wave." Even Merrill Lynch has gotten into the act. In 2004, it issued a small report on nanotechnology and wrote: "We believe nanotechnology could be the next growth innovation, similar in importance to information technology over the past 50 years." The report went on to say that "nanotechnology is real—the questions generally are when, not if."

There is nothing inherently untrue with any of these statements. The only problem is that they will undoubtedly inflate people's expectations over how soon nanotechnology will arrive. The reality is that just as other emerging technologies were marked by both bursts of enthusiasm and then bouts of great cynicism when those initial expectations were not met so, too, will the nanotechnology era. In fact, the field has already experienced one minibubble. In 2004, shortly after the $3.7 billion National Nanotechnology Initiative was passed into law, a number of nanotech stocks soared to new heights in the mistaken belief that many of the publicly traded nanotechnology companies were already doing great things. When the more prosaic truth was learned that most of the companies were years away from producing actual products and generating profits, the stock prices gradually returned to more reasonable levels.

The same thing is likely to happen following the first high-profile nanotech initial public offering (IPO) or Federal Drug Administration (FDA) approval of the first truly nanoscale medical device. Each event will likely be met with a broad, across-the-board increase in the stock prices of many nanotechnology companies—many of which will not be even remotely associated with the companies or technologies in play. This enthusiasm will then be followed by a retraction in nanotech stock prices. (Note: The potential for "momentum" investors to profit from these short-term spikes obviously exists, but because this book is about long-term investing, I'll refrain from saying anything more on this topic.) The best advice is that investors must take all of the hype associated with nanotechnology with a large dose of skepticism.

This is not, however, to say that all hype is bad. Hype helps inspire entrepreneurship. While nanotech will be primarily science and technology-based, it will still require entrepreneurs—risk-takers who understand what it takes to get a business up and operational—to really drive the industry. Jim Von Ehr, founder of Zyvex, a private nanotech company covered in Chapter 7, was originally inspired by the notion of a self-assembler (a device that could essentially create a product from scratch by putting every atom exactly where it needs to go). He has since taken a more practical approach to nanotechnology and has developed real products that have attracted the attention of corporate and government leaders alike. Similarly, Larry Bock, CEO of Nanosys (also covered in Chapter 7 and one of the main contenders to be the industry's first IPO), was inspired by the immense promise of the field. Like Von Ehr, he has taken a practical approach to the industry and is acquiring an impressive portfolio of intellectual property and assessing which products and markets Nanosys can quickly and successfully enter.

Those who criticize hyping nanotechnology now—in 2005—could be compared to those naysayers who said in 1900 that the automobile was being overhyped. It may still be a little premature to hype nanotechnology but to dismiss it entirely is more foolish because it flies in the face of the relentless march of technological progression. The fact is that all revolutionary industries and technologies—from electricity and the transistor to biotechnology and the Internet—all experienced periods of wild-eyed hype as well as busts. All, however, progressed—and so, too, will nanotechnology.

Activity

In the fall 2004, Lux Research, a New York-based research firm specializing in the field of nanotechnology, issued a report in which it forecast that the worldwide market for nanotechnology products and services would reach $2.6 trillion by 2014. If that figure is reached, it will represent close to 15 percent of global manufacturing output.

The $2.6 trillion figure is a sizeable increase over the $1 trillion that the U.S. government has been forecasting as the market for nanotechnology products and services, but Lux was the first firm to seriously attempt to document nanotechnology's impact on the entire value chain and deserves to be taken more seriously than the government's figure—a figure that the federal government has never really explained.

A number of other independent factors give credence to the Lux Research's larger number. The first is the sheer volume of government, corporate, and private money that is being invested in the field. Another indication of how fast the field is maturing is the recent recognition by the U.S. Trade and Patent Office that the field of nanotechnology is mature enough to warrant its own official category—"977." (Such a development might seem insignificant, but given the complexity of nanotechnology-related intellectual property (IP) and legal issues revolving over prior art, patent infringement, and copyright law, IP issues could constitute a serious threat to the future growth of the industry unless addressed early.)

Another significant advance is the work the National Institute of Standards and Technology is doing to define and characterize the standards that will be used in the emerging field. Until scientists, corporations, and customers alike can be sure they are speaking in terms that are widely understood and universally accepted, the opportunity for confusion and disagreements is significant. In fact, in a separate report, Lux Research highlighted how the poor quality of some new nanomaterials was threatening the growth of nanotechnology due to the unreliable nature of the products. The work to codify the standards should help minimize such problems.

The combination of vast amounts of money together with the official accreditation of the field suggests that scores of federal laboratories, academic institutions, corporate labs, and private start-ups doing innovative and exciting work will not only be able to patent and protect their work, but also turn it into commercially viable products.

Nanotech Is Here Now

In June 2004, Wilson Sporting Goods launched its nCode (the "n" is for nanotechnology) tennis racket. The racket is enhanced with carbon nanotubes and is stronger and lighter than graphite frame tennis rackets. Less than three months after its introduction, it was the top selling racket in the high-end market. It is an example of how companies are employing nanotechnology not only to make new products, but also to make existing products better. Within the past year, nanofibers have been placed in over 20 million garments. Nano-Tex, a subsidiary of Burlington, has developed a proprietary technology that makes pants virtually stain resistant. Their Nano-Care technology is now being used by Gap, Eddie Bauer, Perry Ellis, Lee Jean, Old Navy, and Tommy Hilfiger, among others. L'Oreal is using nanoparticles to improve the effectiveness of its lotion, and perhaps even more significant, General Motors is using nanocomposites for stronger and lighter running boards and for the cargo liner of the H2 Hummer. In fact, the entire exterior panel for the new Chevy Impala is constructed out of nanocomposites.

The relentless march of nanotechnology is not going to stop with these advances. As Alan Taub, GM's executive director of global research and development, said of nanotechnology, "It's opening a whole new world for us in the auto industry . . . we're entering a world where we can actually improve on all the critical dimensions rather than make a trade-off." What he is saying is that nanotechnology is going to be able to make next-generation automobiles both lighter and more fuel efficient, and those advances will not require an offsetting compromise in style or passenger safety. In the longer term, nanotechnology will also play a complementary role to GM and other automakers' transition to the emerging era of fuel cell vehicles. From the creation of new materials to store hydrogen to more effective nanoparticle catalysts and to nanofilters for separating and capturing hydrogen, nanotechnology is the enabling technology that will facilitate the transition.

In the computer and semiconductor industries, the importance of nanotechnology underscores IBM's work with carbon nanotubes and self-assembling block polymers—it wants to create smaller, faster, and more powerful computer chips. It explains why Intel Corp. is working with such promising nanotechnology start-ups as Zyvex and Nanosys and why Hewlett-Packard has already purchased some of Molecular Imprints's nanolithography equipment.

In the energy industry, the emergence of nanotechnology explains why Headwaters is attempting to purify coal at the molecular level and liquefy it. ChevronTexaco and Électricité de France, France's largest energy company, have both invested in a promising nanotechnology start-up called Konarka because they understand how its nanotechnology-enabled flexible solar cells could revolutionize energy production. To its credit, Chevron-Texaco has even spun off a nanotechnology company, Molecular Diamonds, to manufacture diamondoid materials in a modest attempt to diversify its business by allowing the company to become a supplier to the semiconductor and pharmaceutical industries. Another start-up, Nanosys, is also working on flexible solar cell technology and has partnered with the giant Japanese conglomerate Matsushita to begin manufacturing products based on its proprietary nanotechnology in 2007.

These energy companies understand that the global demand for energy is going to increase from 13 terawatts of energy per day to 30 terawatts in the next few decades, and if they are to help meet that demand, they must find new and preferably cheap, clean, and renewable sources of energy. One way or another, whether it is clean coal, solar energy, or a new economy based on hydrogen (and fuel cells), nanotechnology is going to play an important part of the future energy equation.

The massive pharmaceutical industry is likewise investing heavily in nanotechnology. Like the semiconductor and energy industries, it understands how nanotechnology will allow it to stay competitive in the short run and position itself for long-term growth. The greatest challenge the industry currently faces is the fact that twenty-three of the world's top drugs will come off patent by 2008. By that time, patents for Lipitor and other leading drugs will have expired, and generic drugs will have been launched to compete with them. That is, unless, the industry can develop new formulations—employing nanoparticles—that allow them to extend that patent or, better yet, create new, improved drugs. Lest readers think such advances are far off, in January 2005, the FDA approved Abraxane, a nanoparticle version of the breast cancer drug Taxol, which is more effective and causes fewer side effects than the leading breast cancer treatment drug.

A Final Word

In many newspapers across the country there is a syndicated column called "News of the Weird." The author usually writes about the follies and exploits of petty criminals and other less than intelligent individuals who get caught in any number of bizarre or ridiculous situations. Occasionally, however, the column ventures off into other areas that the author deems "weird." Once, he noted that researchers at Panasonic's Nanotechnology Research Laboratory in Kyoto, Japan had begun to generate enough electricity from blood to power a battery to run a medical device.

I conclude with this story because far from being "weird," the science is actually quite exciting, and it offers Panasonic (as well as the other companies who are pursuing it) a huge opportunity to enter a massive and lucrative market—namely the multibillion-dollar medical device market—by creating a product that either supplements a battery and allows it to run longer or replaces it altogether.

It is a cautionary tale because in the near future a number of advances might sound "weird" and investors may be inclined to ignore them. My simple advice is: don't. For instance, GE is talking about "printing" lightbulbs, and Hewlett-Packard, as was mentioned earlier, is talking of developing a cell phone "so small an ant could use it." Still others are working on solar cells that can be printed "like wallpaper," seal-healing materials, metal rubber, and molecular-sized "smart bombs" that can kill individual cancer cells.

The markets for these products do not yet exist and as such are impossible for many traditional stock analysts to analyze. But for the individual investor who seeks to understand these advances and can grasp how they might lead to new paradigms or new business models, the future is ripe with opportunity and profit. Chapters 3 through 8 outline many of these companies.

"Fortune favors the prepared mind."
—Louis Pasteur

Chapter 2

Due Diligence

In the early 1980s it was common for companies to add the term "micro" to their name in an attempt to give them some cache. In the late 1990s, the business world experienced a similar phenomenon when companies rushed to recast themselves as dot.coms. The early part of the twenty-first century will likely see the same thing with the term "nano."

In fact, it has already started. A small company, which had previously gone by the nondescriptive name Mendell-Denver (and before that Sunlight Systems), changed its name a few years ago to Nano-Pierce Technologies. The company, in spite of statements and press releases to the contrary, has nothing to do with nanotechnology. Still, in the past year, its stock price has fluctuated greatly, and it reached its peak at the same time the National Nanotechnology Initiative was passed into law. Amazingly, for a company with no revenues, it reached a market capitalization of $134 million.

Another company, NanoMetrics, must also be viewed with a jaundiced eye—but for a difference reason—its stock ticker symbol is "NANO." In spite of this, the company is not really involved in nanotechnology—and to its credit the company does not claim to be—but in early 2005 when it announced it was merging with another company, August Technologies, its stock jumped on the news. Some momentum investors seeing an opportunity to make some easy money pumped up the stock as a "nanotechnology pure play" on various web-related bulletin boards, and its price increased nearly 10 percent in a day before falling back to less than its original price when the reality that the "news" was not related to nanotechnology but rather just a merger of two small semiconductor companies became evident.

These two stories underscore the need for investors to do their own due diligence prior to investing in any nanotechnology company. In both cases, investors could have prevented being on the short end of the stick if they had done just a little homework. As the field of nanotechnology grows and as more investors get involved in the area, such stories will only become more common. What follows then is a list of questions that should be asked—and answered—prior to investing in any nanotechnology company.

Strip the "Nano" Label

The first step any individual investor needs to take when conducting due diligence is to strip the term "nanotechnology" off whatever the company is doing and investigate it from a standard business perspective. The general rule of thumb is to invest in good business opportunities, not in broad definitions like "nanotechnology." While nanotechnology can—and will—give certain companies a real advantage, there are simply too many companies using the term too loosely for investors to take any company's word at face value.

Ironically, if the company does not purport to be a nanotechnology company, that is, more often than not, a positive sign. For instance, ZettaCore, one of nanotechnology's most promising start-up companies, refuses to even call itself a nanotechnology company. The company's CEO goes out of his way to state that ZettaCore is a "semiconductor" company that just happens to be employing nanotechnology.

The distinction between NanoPierce and ZettaCore leads to the first set of questions any individual investor must consider before investing in any nanotechnology company.

Does the company talk about specific market applications for its technology or just large markets? ZettaCore talks in clear terms about how its technology can create high-density data storage devices. NanoPierce, on the other hand, only talks in vague terms about the multibillion dollar computer industry without ever explaining how it intends to be a player in the field. *Beware of any company that throws around big numbers and claims its products will capture a sizeable share of any multibillion-dollar industry.*

Companies should be able to subcategorize the specific market they intend to enter (e.g., electronics, tools, biotechnology). Companies that claim to be applying nanotechnology to a wide range of markets need to be treated with suspicion. For instance, NanoPierce also claims to have created a yeast product that will take the "$12 billion poultry industry" in Georgia by storm. Just why a self-proclaimed nanoelectronics company is also looking at the poultry market is never elaborated on by the company's representatives, but the wide divergence of the two markets should again give any investor cause for concern.

Does the company talk about product development within a reasonable time frame? Better yet, has it actually produced a real product? Companies that are only in the concept or development stage are probably still at too early a stage for the average individual investor to invest his money.

Finally, does the company have strategic partners or actual customers? Many of the markets that nanotechnology-enabled solutions will find a home in—semiconductors, pharmaceuticals, medical diagnostics and devices—are large and complex markets. As such, they are difficult for small companies to crack. Having a strategic partner is often the best, easiest, and fastest way to commercial success.

In addition to these questions, there are four factors individual investors should also take into consideration. These four factors can be thought of broadly as people, markets, technology, and finances. The questions to consider are as follows: (1) Does the company have a reputable and experienced management team? (2) Can its product or technology be mass produced quickly, cheaply, and reliably? (3) Does the company possess technical leadership in its field, and does it have propriety intellectual property? (4) Does it have the financial resources to accomplish its strategic goals?

It All Starts with People

Obviously, it is not wise to focus on just one of these four factors. They have to be viewed together as part of a whole picture. However, when beginning one's due diligence, a lot of time can be saved by researching the quality of the management team. The quality of a company's management has the highest correlation to whether it succeeds or fails. An innovative or "cool" technology is not enough to guarantee success. An experienced CEO is often necessary to drive the right technology solution to the largest market. Furthermore, because it is rare that any technology—or business—ever evolves according to plan, an executive team that has actual experience growing a business is a definitive advantage. Often, these executives have learned from past mistakes and will have developed the capacity to adapt to rapidly changing environments.

Many nanotechnology companies are likely to be started by scientists. Investors should not be lulled into believing that their scientific credentials alone provide them with the skills to run a company. These scientists are often brilliant and understand their technology better than anyone else. They are not, however, managers or executives. Scientists don't always understand the marketplace. Moreover, they aren't trained to take risks—scientists are taught to be methodical. The trait is a necessary ingredient in science, but it can be deadly in business—especially in a business environment that is changing as rapidly and radically as nanotechnology. Good executives know when to act, and they often need to do so with less than perfect information.

Other "people" factors investors will want to consider include assessing the scientific diversity of the executive team. The truly exciting opportunities in the field of nanotechnology are going to come from the convergence of electronics, biology, physics, and chemistry at the nanoscale. The more extensive the company management and technical team's broad scientific perspective, the more likely it is they will be able spot these emerging opportunities or take advantage of the natural synergies that nanotechnology often offers.

Related to this point, potential investors are encouraged to review the scientific advisory board the company has assembled. Does it have the depth and breadth of experience to really direct the company? Are the advisors really part of the management team, or are they "paper only" members. The more engaged they are in the company—the better.

It's the Product

In the 1930s, it was demonstrated that a new keyboard, called the Dvorak system, was superior to today's common QWERTY keyboard. It allowed skilled typists to type an average of 165 words per minute versus 131 words on the QWERTY system. It did this by rearranging the letters so that there was less left hand use, fewer row-to-row hops, and none of those bothersome pinky stretches.

As history has vividly demonstrated, the Dvorak system, in spite of its superiority, didn't win in the marketplace. The reason is because it required people to learn an entirely new system of typing. While it would undoubtedly have been more efficient for those doing a lot of typing, for most users it did not justify the up-front investment in time to learn a new system that would yield only a modest increase in efficiency.

The moral of this story is that just because a new technology is better does not guarantee that it will win in the marketplace. It also serves as a cautionary tale about the difficulty of assessing whether a market will embrace a new technology. Normally, if you told a consumer or a company that a product that would yield a 20 percentage point increase in efficiency, they'd jump at it. Such is not the case if it requires the consumer to change behavior.

This is relevant for a variety of products and technologies that will either be improved or enabled by nanotechnology.

Many of the improvements will be expensive, require new support systems, or simply may not offer enough of an improvement to justify a change.

To help determine whether a technology or nanotechnology-enabled product has "legs," investors should be able to answer the following questions: Does the product solve a real problem for its customers? For instance, does it save its user time, improve their health, or provide them a freedom they didn't previously enjoy? If the product meets a real need, then investors have something worth considering. If not, investors should consider leaving it for others to fund.

A related question is: Does the product require a change in user behavior—like the Drovak system did. If so, the challenge is significantly greater. NEC is developing a new cell phone that derives its energy from a fuel cell using carbon nanohorns. The fuel cell will reportedly last longer than an average battery, but it will also require users to "refill" their phone. It remains to be seen how consumers will respond to the change, but it is something investors need to consider before investing.

Is It an Idea, a Demo, or a Real Product?

The next step in discerning the viability of a product is to assess the development stage of the company's product. For example, is it in the concept, preproduction, or postproduction stage? If it hasn't matured to the point where it is past the concept and an actual prototype has been developed, it is too early for most investors.

If a company's technology or product is past the concept stage, the next question investors need to consider is whether the company has demonstrated "scalability and reliability." Can the nanomaterials or nanodevice in question be manufactured in the quantities and sizes necessary to attract the attention of major customers? Can those products be made in a manner consistent enough to guarantee quality and performance? A number of companies have recently begun producing carbon nanotubes that have a host of amazing properties, including a high strength-to-weight ratio and conductivity. As a result, their use is being explored in everything from flat panel displays to drug delivery devices. The problem is that carbon nanotubes are not easily manufactured with the exact properties the end user may want. Until this happens, every company in the field remains a question mark.

The most promising sign that a nanotechnology company is on the verge of creating a viable business—at least in the short term—is if its technology does not require manufacturers to change any of their existing equipment or processes. As was stated earlier, it is human nature to resist change, and large companies are no different. Those companies that create technologies that don't require manufacturers to change are going to have a leg up, at least, in the short term. An example of this is Nantero's carbon nanotube-based technology. The company claims its technology will not require existing computer memory companies to change anything. In fact, LSI Logic has indicated it will try the technology in fall 2005. If successful, LSI's approval would validate Nantero's technology and likely attract the attention of other companies.

Regardless of where the product is in the development stage, investors should determine whether the company has done its homework in regard to how it's going to approach the marketplace. Does the company demonstrate pricing logic? Has it determined why a customer would be willing to pay a specific price? Has it targeted specific customers? Better yet, does it already have customers?

One Is the Loneliest Number

Because many nanotechnology start-ups are small, they will need assistance in getting their product to market. For this they will often need partners. Therefore, at a minimum, investors should know whether a company has successfully entered into arrangements with large corporate partners who will either use their products or help produce their products. For instance, Zyvex, a promising private nanotechnology company, is working with Intel to validate whether its carbon nanotubes may be used to create a better thermal interface for the next generation of computer chips (the material may help alleviate the problem of overheating). Another excellent example of good partnering is Nanosys's arrangement with Matsushita to begin producing and distributing flexible plastic solar cells by 2007. On its own, it would just be a promising technology. With Matsushita's backing, it is a formidable threat.

Such an arrangement need not, however, be exclusively with large corporations. In late 2004, Advanced Micro Devices announced it was teaming up with Albany Nanotech (New York state's huge $100 million public-private nanoelectronics program) to find ways to improve strained silicon—a material that is already being used to improve the performance of today's state-of-the-art integrated circuits—and will likely lead to even more improvements.

A few other product-related questions to ask before investing include:

- ► Can the company's product evolve over time?
- ► How will the product be marketed and sold?
- ► Does the company have access to foreign markets? If it is a foreign company, does it have access to the U.S. market?

Beware of Competing Technologies ... and Lawyers

As the two previous sections on people and markets demonstrate, investors cannot rely on superior technology alone to drive a company's stock upward. Having said this, technology is still obviously important, and assessing a company's technology—and the intellectual property behind it—is among the most difficult and time-consuming tasks.

Nanotechnology, by its very nature, is incredibly complex and requires a broad base of scientific knowledge. Specifically, it requires a deep understanding of many different fields of science—biology, physics, chemistry, material sciences, and the computational sciences. For the average investor (and even most professional investment advisors), assessing the relative merits of technology enabled by nanotechnology is beyond their skill set.

How then does one go about it? The most important thing to understand is what other technologies (and companies) are out there trying to address the same problem. For instance, in Chapter 3 there are three companies developing nanolithography imprint equipment. Chapter 4 will list a host of companies developing nanoparticle catalysts, and Chapter 7 will document a handful of companies developing flexible solar cells. The best advice is to beware of these competitors and then let the companies themselves explain why their technology is superior.

The one thing investors should not do is be overly impressed with the number of patents a company has. All patents do is exclude others from practicing the invention. It does not stop someone from creating a different way to address or solve the same problem. One company can hold 250 worthless patents, while another can possess just one very valuable patent.

The rub lies in distinguishing a worthless patent from a valuable one. Recognizing that this skill is also beyond the capability of most people—at least in a field like nanotechnology—investors are encouraged to look at the scientific credentials of the founders of the company and its scientific advisory board. This is by no means a perfect measure, but to the extent that the individuals associated with the company have published papers in credible, peer-reviewed scientific journals or have established relationships with credible academic institutions, government laboratories, or corporations, it is a positive sign.

For instance, the fact that Richard Smalley, founder of Carbon Nanotechnologies, is also the 1996 Nobel Prize winner in chemistry doesn't necessarily guarantee his patents or related technology relating to carbon nanotubes are superior to others; however, it does improve the odds the company's technology is on solid ground. Similarly, the fact that Charles Lieber and Paul Alivisatos—both of whom are involved with Nanosys—have had a number of articles published on their nanotechnology research and are widely recognized as world-class researchers in their respective fields suggests that Nanosys's technology and intellectual property warrants attention.

Investors will also want to take into consideration whether the smaller nanotechnology companies are partnering with large corporate companies or have received investments from leading venture capital firms. The reason is because both have scientists and trained technical advisors with the requisite skills to more thoroughly evaluate a company's technology and its intellectual property. All things being equal, if established companies and venture capital firms have assessed the technology and decided to invest in the company, it is a positive sign.

The tactic essentially amounts to letting others do your due diligence for you, but unless one has the technical skills and the time, it is often the best that can be done. Chapter 8 lists a few of the venture capital firms who have developed some expertise in nanotechnology, and in the company profiles in the following chapters every attempt has been made to list which venture capital firms and large companies have invested in a given company or are partnering with the company. The information should be considered a proxy for the viability of a company's technology.

Such measures are imperfect, but they pale in comparison to the difficulty of assessing intellectual property. Furthermore, it is almost a given that any successful technology will draw some type of legal challenge, and that challenge will come only after time, money, and a great deal of effort has already been invested into getting the technology to the marketplace.

The best way to assess a company's position in this regard is to determine if the company itself has done its own due diligence on its intellectual property. Questions to ask are:

- ► Has the company thoroughly analyzed its own IP claims?
- ► Has it analyzed the patents held by its competitors?
- ► Does it have international patent protection?
- ► Does it have systems in place to protect its IP?

If a company has licensed its intellectual property to others, investors should understand:

- ► The terms and conditions of the license. Is it an exclusive, nonexclusive, or field-of-use exclusive license?
- ► What is the duration of the license?
- ► How is the patent holder compensated—in cash, equity, royalties, or some combination thereof?
- ► If there is a challenge, who is responsible for paying the patent prosecution costs?

As with the assessment of the technology itself, assessing such legal issues is best left to the experts—in this case, the lawyers. Because such expertise is beyond the financial means of the average investor, we are again left with the situation of relying on the legal experts of the company, partnering companies, or venture capitalists. Often, the best an investor can do is ask the questions, and if the answers are not satisfactory, if there are too many unanswered questions, or if it appears a legal challenge could either delay or entirely stop the successful introduction of the technology, it is best to hold off on an investment until those issues are resolved.

Many companies, even private companies, often have a staff person devoted to investor relations. Investors are encouraged to contact these individuals and seek answers to the preceding questions.

Follow the Money

In real estate, realtors are fond of saying that the three most important things are location, location, and location. Some in the investment field have parroted this line and said that the three most important things for any new business are money, money, and money.

Money is obviously an important component of any business, and no business can succeed without it. For established businesses, profits are an absolute necessity over the long run. But for start-ups, the situation is a little more complex.

It is unwise to give too much attention to how much money a new private start-up has raised. For one thing, too much money can be a bad thing in the sense that it can result in an undisciplined business atmosphere where company executives and employees don't feel a need to squeeze out every efficiency. It may also allow company executives in the short to midterm to cover over—and hide from investors—some fundamental problems.

When conducting due diligence on start-ups, there are a few keys factors one should consider. The first is to remember that it is unwise to fund a research project. More simply put, investors should only consider investing in those companies that have moved beyond the "idea" stage and are actually manufacturing—or are close to manufacturing—products. As said earlier, the manufacturing process should be mature enough that products can be built on a reliable, cost-effective, and scalable basis.

The second thing to look for is something called the "skin game." Do the company founders have their own money invested in the company? Even more important perhaps is whether they have convinced their family and friends to invest in their company? If the answer is yes to both questions, it is a positive sign. It speaks to the founders' confidence in the company, and it provides them a stronger incentive to succeed—no one likes to let down their family or friends.

The third factor to look for is government money. Investors should not fund research or concept stage projects, but governments should—and often do. Therefore, investors are encouraged to consider whether a company has received grants from NASA, the Department of Defense, or the National Institutes of Health—to name a few of the more popular government agencies funding nanotechnology projects. Almost every company listed in Chapter 7 has received some government funding. In fact, a few have received very sizeable grants. For instance, Zyvex received a $25 million grant from the National Institute of Standards and Technology (NIST) to fund the development of equipment that can manipulate matter at the atomic level; while

Nanosys and Konarka have received multimillion dollar grants to employ nanoscale materials in the development of flexible plastic solar cells.

The point here is not to imply that the government has an impressive track record at picking winning technologies, rather it is to highlight the fact that the government is, in essence, helping to underwrite a company's research and development—and it is doing it in a way that doesn't dilute investor equity. (The government doesn't ask for a stake in the company—only a right to use the technology if and when it is developed.) Investors should, however, be cautious of companies that are either entirely reliant on government grants or who, after years of receiving government funding, have been unable to attract any corporate attention.

Investors are also encouraged to consider the amount of venture capital funding a company has received. This is a dual edge sword. On the positive side of the ledger is the fact that these firms have done their own due diligence and have found enough promise in the company to warrant an investment. Not all venture capital firms are equal, however. As was demonstrated in the dot.com era, a herd-like mentality can often be found among venture capital companies. At the present time, only a handful of firms have acquired the expertise to adequately do the due diligence in the field of nanotechnology. These firms are reviewed in Chapter 8.

In addition to the financial resources they bring to the table, venture capital firms are important for two additional reasons. First, they often come to the table with "fat Rolodexes" and can help the companies they invest in find the appropriate executive management team. Secondly, the good venture capital companies have existing relationships with major corporations and can use those relationships to play the role of matchmaker.

The downside is that for assuming so much risk venture capital firms often demand a sizable share of the company's equity. This is a dilemma for both the company founders and individual investors. Obviously, venture capital firms deserve to be rewarded for the risk they assume, but the question is how much. There is simply no easy rule to follow. The more money and the earlier it invests, along with the amount of scientific and professional assistance it brings to the table, all need to be considered.

In the final analysis, venture capital is usually a positive thing. Most start-ups fail—even those in which venture capitalists invest. Venture capitalists help fund the development of the idea, professionalize the management, and assist the company in getting its product to the right market in a time frame that allows the company the best chance of succeeding.

Buyer Beware

The harsh reality of the marketplace is that most high technology companies fail. Nanotechnology is not going to be any different, and various nanotechnology companies will fail for the same reasons other companies do: poor management, inferior technology, and undercapitalization. (A host of other factors are also at play, and these will be discussed in greater detail in Chapter 9).

By doing due diligence, however, the individual investor can reduce their risk. Let it be stressed, however, that while risk can be reduced it cannot be *eliminated!*

Doing due diligence is not an easy task, but here are the ten most important questions an investor should have answered before investing in any company.

1. Is the company's management team experienced?
2. Does the company's product meet a real-world need?
3. Is the product ready for the marketplace, and can it be produced consistently and reliably?
4. Does the company have strategic partners?
5. Does the company's founder have a strong scientific and technical background?
6. Is the company's board of scientific advisors actively engaged in the company?
7. Is the company's intellectual property patented, or has it secured the necessary licensing agreements on favorable terms (e.g., exclusivity, duration)?
8. Do the company's founders have their own money invested in the company—and that of their family and friends?
9. Has the company received any government grants to help fund its research and development?
10. Has the company received venture capital from a firm with established expertise in the area of nanotechnology?

Louis Pasteur once said, "Fortune favors the prepared mind." This is especially true when investing in a new, emerging, and fast paced field like nanotechnology. Those who are most likely to profit are those who have taken the time to assess the technology, a company's management, and the competitive marketplace—for they are most likely to avoid the pitfalls and spot the opportunities.

"Let me be as clear as possible: if the Internet improved our quality of life via the Information Superhighway, then nanotechnology should be considered the Express Lane for future technological breakthroughs to make our lives simpler, safer and more enjoyable."

—Jim O'Connor, Vice President of Technological Commercialization

Chapter 3

The First Winners: The Nanotechnology Enablers

Willy Sutton, an infamous thief in the 1930s, was once asked why he robbed banks. His response has become equally infamous: "Because that's where the money is." A similar thought is at work for nanotechnology investors. Long before investors can hope to reap the profits from the $1 to $3 trillion in nanotechnology products that are expected within a decade's time, those products and devices must first be manufactured. To do so, very specialized equipment will be necessary.

Therefore, one of the safer strategies for investing in nanotechnology is to seek out investments in the companies who are manufacturing the equipment that will enable the nanotechnology revolution because that, at least in the short term, is "where the money is." In 2006, it is expected that governments, corporations, and smaller nanotechnology start-ups will invest close to $10 billion in nanotechnology research and development. A large portion of these investments will go to purchase the equipment and software that can visualize, characterize, and manipulate matter at the atomic and subatomic level.

In the investing world, the strategy of investing in the underlying equipment that will enable a new, promising field is known as the "pick and shovel" approach. It is a reference to the Gold Rush of 1849. Thousands of prospectors went out West with the hope of finding gold and striking it rich. As history has recorded, only a few succeeded. Among those first to get rich weren't the panhandlers, they were the businesses that catered to the thousands of speculators by selling them the picks and shovels they needed to ply their trade. Among the more famous was Levi Strauss, who made a sizeable fortune selling durable denim jeans.

A more recent analogy is the Human Genome Project. Started in 1994, the U.S. government and Celera, a private corporation, began an expensive race to sequence the human genome. The government, through the National Genome Research Institute, invested an estimated 10 percent of its total funding in Applied BioSystems, who manufactured the equipment necessary for sequencing the genome. As a result, the company's stock soared in the late 1990s.

With the signing of the National Nanotechnology Initiative in December 2003, the U.S. government has now committed to spending roughly $1 billion per year for the next four years on nanotechnology-related research and development. The government and university labs funded to conduct this research will need scanning electron microscopes (SEM), atomic force microscopes (AFM), transmission electron microscopes (TEM), and a variety of other tools and software to conduct their research. It stands to reason, therefore, that among the first winners in the field of nanotechnology will be those companies who are creating the necessary tools. This approach to investing, while it does not promise the dramatic returns of some of the companies offered in Chapters 6 and 7, does have the benefit of being less risky because these companies are actually producing real products today.

What follows is an overview of those publicly traded companies that are manufacturing some of the more commonly used equipment and software in the field of nanotechnology today. A short section then follows documenting some of the more promising private nanotechnology start-ups in this area who may either go public soon, become acquisition targets for the larger nanotechnology equipment companies, or who are working on technology so different that, if accepted by the marketplace, it could end up taking significant market share away from some of the publicly listed companies.

To aid the reader, each company's contact information is provided, along with a brief description of the company. For publicly traded companies, readers are provided bullet points listing both the positive (i.e., "Bullish") reasons for investing in the company as well as the negative (i.e., "Bearish") reasons. For each company, a personal opinion for the company's prospects is rendered and concludes with a short section entitled "What to Watch For" that will provide some guidance to assist you in tracking future developments.

Readers are also encouraged to do an accompanying Internet search before investing in any nanotechnology company to apprise themselves of developments in this fast-moving field that may have occurred since the publication of this book. (A list of resources is included in Chapter 9.)

ASMI	COMPANY	ASM International
	SYMBOL	ASMI
	TRADING MARKET	NASDAQ
	ADDRESS	Jan van Eycklaan 10 Bilthoven, The Netherlands 3723
	PHONE	212-686-8144
	CEO	Arthur Del Prado
	WEB	*www.asm.com*

DESCRIPTION   ASM International designs, manufactures, and sells equipment used to produce semiconductor devices including atomic layer deposition equipment and thermal processing systems capable of forming junctions and contacts in the sub-45 nanometer range. It is currently included in the Activest Lux Nanotech Mutual Fund.

REASONS TO BE BULLISH

► In 2004, ASM's revenue grew 41 percent, and the company was profitable.
► The company's strong relationship with IMEC, Europe's leading nanoelectronics and nanotechnology research center, and its R & D budget suggests it will stay abreast of next-generation developments and bodes well for future growth.

REASONS TO BE BEARISH

► The industry is cyclical and very competitive. ASM faces competition from KLA-Tencor and Applied Materials.
► The weak dollar could hurt ASM's growth prospects by making its products less competitive in the United States.

WHAT TO WATCH FOR   The successful transition to nanoimprint lithography could seriously erode ASM's market. The advances of Obducat, Molecular Imprints, and others pursuing nanoimprint lithography technology should be tracked closely.

CONCLUSION   Outlook is bullish. ASM's continued growth in 2005 and its prospects for continued growth in the midterm (2006–2007) make it an attractive stock. In the long term, the company's relationship with IMEC should help it stay competitive as the semiconductor industry moves to the sub-45 nanometer range.

Publicly Traded "Pick 'n Shovel" Nanotechnology Companies

ACCL	COMPANY	Accelrys, Inc.
	SYMBOL	ACCL
	TRADING MARKET	NASDAQ
	ADDRESS	9685 Scranton Road San Diego, CA 92121
	PHONE	858-799-5000
	CEO	Mark J. Emkjer
	WEB	*www.accelrys.com*

DESCRIPTION   Accelrys develops and licenses molecular modeling and simulation software to the chemical and life sciences industries. Its software is used by biologists, chemists, and material scientists for product design and drug discovery. In 2004, Accelrys spun off from Pharmacopeia in order to focus exclusively on molecular modeling and simulation.

REASONS TO BE BULLISH

► In the first quarter of 2005, revenues were up 88 percent over the same period the previous year.

► Accelrys counts some of the world's leading chemical and pharmaceutical firms, such as Dupont, Dow Chemical, BP, Eli Lilly, Pfizer, Amgen, and Genencor, among its customers.

► It has started two strategic partnerships, one with IBM to develop modeling and simulation tools for the Linux operating systems (which is expected to provide pharmaceutical companies more flexibility and increased capabilities to design new drugs) and the other with Sigma-Aldrich to integrate its chemical compound catalogs into Accelrys's database.

► Accelrys has excellent working relationships with academic institutions, including the University of Cambridge and Harvard, as well as the top government research laboratories involved in nanotechnology such as Argonne, Brookhaven, and Los Alamos, suggesting that its software is useful in cutting-edge research.

continued

Accelrys, Inc. continued

► It has also initiated the Accelrys Nanotechnology Consortium—a program designed to accelerate the development of software tools that enable the design of nanomaterials and nanodevices. Current members include Corning, Fujitsu, e2v Technologies, Imperial College, Uppsala University (Sweden), Franhofer, and the Japan Advanced Institute of Science and Technology.

REASONS TO BE BEARISH

► In 2004, Accerlys reported a loss of $29 million on revenues of just over $84 million.
► It faces strong competition in the pharmaceutical arena from Tripos (TRPS), whose customers include AstraZeneca, Pfizer, Bayer, Hewlett-Packard, and IBM Life Sciences, as well as from NanoTitan, a small nanotech software modeling firm located in McLean, Virginia.

WHAT TO WATCH FOR   If leading life sciences firms begin switching to the Tripos software or to private start-ups such as Apex Nanotechnologies or NanoTitan, investors will want to review their holdings in Accelrys. The company is also attempting to move into the "services" market. If it is successful, it could yield some additional revenues.

CONCLUSION   Outlook is bullish. Software tools are going to be essential for nanotechnology research—be it at the academic or corporate level. The need to view, characterize, and understand materials at the nanoscale will only increase in the future. As the need to perform calculations increases from hundreds to thousands of atoms (and beyond), the software will need to keep pace. Provided Accelrys stays at the cutting edge, it should be well positioned for future growth.

Publicly Traded "Pick 'n Shovel" Nanotechnology Companies

FEIC		
	COMPANY	FEI company
	SYMBOL	FEIC
	TRADING MARKET	NASDAQ
	ADDRESS	5350 NE Dawson Creek Dr. Hillsboro, OR 97124
	PHONE	503-726-7500
	CEO	Vahe A. Sarkissian
	WEB	*www.feicompany.com*

DESCRIPTION   FEI manufactures and distributes focused ion beam systems (FIB), scanning electron microscopes (SEM), and dual beam systems (a combination of the SEM and FIB) that aid in the assessment of 3-D structures at the atomic and sub-atomic level. In 2004, FEI announced its 200kV transmission electron microscope had viewed images with a resolution of less than one angstrom (an angstrom is one-tenth of a nanometer.) The FIB works like an atomic sandblaster to cleanly remove material from a surface. It can also be used to repair defects on photomasks or rewire an integrated circuit by depositing a wire on an integrated circuit.

REASONS TO BE BULLISH

► FEI has been profitable since 2000, and in 2005, it introduced a new product, The Titan, which should continue to increase revenues. In the first quarter of 2005, revenues were up 15 percent over the same period the previous year.

► Current customers include Intel, Seagate, Hewlett-Packard, the University of California-Berkeley, and Stanford, and its sales are well diversified across the globe with roughly one-third coming from the United States, one-third from Europe, and one-third from Asia.

► FEI recently developed a new center in Europe, called Nanoport, to work on joint research and development projects, including the European Union's Interaction Proteome project.

► It has developed a new product—the CLM 3D—that is capable of generating a lot of new orders.

continued

FEI Company continued

REASONS TO BE BEARISH

► FEI's sales are sensitive to the overall health of the semiconductor industry. To the degree that the semiconductor industry is in an economic down cycle, FEI will be adversely affected.

► The company faces increased competition and market pricing pressure. KLA-Tencor, JEOL, NEC, SII Nanotechnology, and Rave LLC are all eager to enter the market, and a shakeout in the mask repair business is quite possible, as too many vendors are chasing after too few dollars in the arena.

WHAT TO WATCH FOR   In the past, FEI's gross margins have not matched their industry peers. If management does not get the problem under control, it will be a cause for concern. Also as the semiconductor industry moves to the sub-65 nanometer level, investors need to keep an eye on whether nanoimprint lithography takes hold in the industry. If sales of Molecular Imprint, NanoNex, or Obducat's nanoimprint equipment rise significantly, investors will want to reconsider their FEI holdings.

CONCLUSION   Outlook is bullish. FEI is a solid long-term investment. As the semiconductor industry continues to progress to the 90, 65, and 45 nanometer range, it will have an increased need to visualize at those levels. FEI's tool can operate at these levels. Its equipment should continue to help customers reduce process development costs, improve yields, and reduce time-to-market. Moreover, because its equipment can view images at the subnanometer range, the life sciences sector will likely use its equipment to help view and characterize viruses, cells, DNA, and proteins. Specifically, FEI's 3-D electron microscopes will help scientists understand the chemical basis for life and uncovering the mechanisms of disease.

6951.JP		
	COMPANY	JEOL
	SYMBOL	6951.JP
	TRADING MARKET	Toyko Stock Exchange
	ADDRESS	(U.S. Office) 11 Dearborn Road Peabody, MA 01960
	PHONE	978-535-5900
	CEO	Robert T. Santorelli
	WEB	*www.jeol.com*

DESCRIPTION   JEOL is a manufacturer of scanning electron microscopes (SEMs), scanning probe microscopes (SPMs), transmission electron microscopes (TEMs), and a variety of nanotechnology-related pieces of equipment. It is one of eleven companies that are included in the Activest Lux Nanotech Mutual Fund.

REASONS TO BE BULLISH

► The Japanese government, like the U.S. government, is investing heavily in nanotechnology research and development, and JEOL, as the leading Japanese manufacturer of nanotech-related equipment, will likely benefit.

► Many academic institutions and leading private start-up companies are currently using its equipment.

► JEOL also manufacturers mass spectrometers (which measure the weight of molecules) and nuclear magnetic resonance equipment (which identifies molecules) and thus has a more diversified product line than some of it competitors.

REASONS TO BE BEARISH

► JEOL is competing against Veeco and FEI and does not have as strong a presence in the United States.

WHAT TO WATCH FOR   Keep an eye on what equipment the private start-up nanotechnology companies are buying—especially the truly "disruptive" companies covered in Chapter 7. If they are buying JEOL equipment, it is a bullish sign.

CONCLUSION   Outlook is bullish. JEOL's reputation is excellent, and it is likely to continue to experience growth. Investors interested in a conservative bet on nanotechnology should consider an investment.

JMAR	COMPANY	JMAR Technologies, Inc.
	SYMBOL	JMAR
	TRADING MARKET	NASDAQ
	ADDRESS	5800 Armada Drive Carlsbad, CA 92008
	PHONE	760-602-3292
	CEO	Ron Walrod
	WEB	*www.jmar.com*

DESCRIPTION JMAR is a manufacturer of X-ray lithography steppers that are a potentially important component for manufacturing next-generation (sub-45 nm) integrated circuits. It also manufacturers collimated plasma lithography (CPL) technology that can be used for the fabrication of high-speed gallium arsenide semiconductors.

REASONS TO BE BULLISH

▶ JMAR continues to receive government grants—particularly from the U.S. Department of Defense and DARPA—which suggests its technology has defense-related applications.

▶ As the demand for advanced high-speed communication devices such as cell phones and digital assistants (which rely on gallium arsenide semiconductors) grow, the demand for CPL technology may also grow.

REASONS TO BE BEARISH

▶ JMAR's revenues fell 30 percent in 2004, and its net loss increased from $2.3 million to $5.3 million. The trend continued in the first quarter of 2005.

▶ The U.S. government remains its largest customer, and it has no major industrial, commercial partners or customers.

WHAT TO WATCH FOR If the company can demonstrate that there is a demand for its products in the commercial marketplace, investors will want to start tracking the stock.

CONCLUSION Outlook is bearish. Due to its declining revenues, increasing losses, and the lack of private sector customers, JMAR remains a high risk. Investors would be better off taking a pass and allowing the government to fund JMAR's research and development until some commercial users are found.

> **Publicly Traded "Pick 'n Shovel" Nanotechnology Companies**

KLAC		
	COMPANY	KLA-Tencor
	SYMBOL	KLAC
	TRADING MARKET	NASDAQ
	ADDRESS	160 Rio Robles San Jose, CA 95134
	PHONE	408-875-3000
	CEO	Kenneth L. Schroeder
	WEB	*www.kla-tencor.com*

DESCRIPTION   KLA-Tencor is the world's sixth largest computer chip equipment supplier and is the largest supplier of process diagnostic and control (PDC) equipment. KLA-Tencor's equipment saves chipmakers money by reducing the number of defective semiconductors produced.

REASONS TO BE BULLISH

► Through the third quarter of FY2005, net income had risen 86 percent from $66 million to $123 million.

► Chipmakers will continue to need to make strategic investments in process control equipment to develop and manufacture next-generation chips, and KLA-Tencor, as the market leader in PDC equipment, is well positioned for future growth.

► The company's partnership with IMEC, Europe's leading nanoelectronics and nanotechnology research center, to accelerate the adoption of next-generation metrology technology for 65 nm and below semiconductor applications also bodes well for long-term growth.

► In the summer 2004, KLA announced a strategic partnership with SII Nanotechnology (a subsidiary of Seiko Instruments) to distribute SII's new atomic force profiler—the Nanopics 2100—which appears to have significant advantages over many traditional atomic force microscopes.

► The fact that it has invested heavily in research and development and has taken an equity stake in Molecular Imprints (see page 68) suggests it has its eye on even longer-term developments.

continued

KLA-Tencor continued

REASONS TO BE BEARISH

► KLA-Tencor is subject to the deep cyclical swings of the semiconductor industry.

► It faces tough competition from larger companies like Applied Materials.

WHAT TO WATCH FOR   KLA-Tencor needs to continue to invest heavily in R & D. If the company misses a technology cycle (i.e., if a competitor produces equipment that has superior diagnostic capability at the 65 or 45 nm range), the company could be seriously damaged.

CONCLUSION   Outlook is bullish. KLA-Tencor is a well managed company with a significant market position in an area—PDC equipment—that is expected to grow. It promises solid—albeit not spectacular gains—for the foreseeable future.

MFIC.OB	COMPANY	MFIC Corporation
	SYMBOL	MFIC.OB
	TRADING MARKET	Over-the-counter
	ADDRESS	30 Ossipee Road Newton, MA 02464
	PHONE	617-969-5452
	CEO	Robert P. Bruno
	WEB	*www.mficcorp.com*

DESCRIPTION   MFIC manufactures submicron materials processing equipment. Its nanoparticle manufacturing systems can be used in biotechnology, pharmaceutical, chemical, cosmetic, and food processing industries.

REASONS TO BE BULLISH

► MFIC's revenues grew slightly in 2004 to $12.2 million, and the company posted a modest profit of $745,000. In the first part of 2005, revenues continued to increase slightly.

REASONS TO BE BEARISH

► MFIC's nanoparticle and carbon nanotube (CNT) production systems have not yet achieved any significant market penetration.

► The company has little working capital, suggesting it may not have the resources to achieve the necessary size to become competitive.

WHAT TO WATCH FOR   The company is reportedly working on developing equipment that can arrange carbon nanotubes in a specified manner. If it were to achieve this capability, it would be a positive sign, and the leading developers of CNT's would likely purchase this equipment.

CONCLUSION   Outlook is bearish. MFIC is a small company in a very competitive market. Until it can achieve greater market penetration, investors should stay away from its stock.

MTSC	COMPANY	MTS Systems Corporation
	SYMBOL	MTSC
	TRADING MARKET	NASDAQ
	ADDRESS	14000 Technology Drive Eden Prairie, MN 55344
	PHONE	952-937-4000
	CEO	Sidney Emery
	WEB	*www.mts.com*

DESCRIPTION   MTS develops and manufactures a series of products that help determine the mechanical behavior (such as wear, thickness, hardness, and elasticity) of materials, products, and structures including nanomaterials. It is one of the twenty-six companies listed on Merrill Lynch's Nanotech Index.

REASONS TO BE BULLISH
► In 2004, the company's revenues increased 8 percent, and it had a profit of $29 million. The trend continued in the first part of 2005.
► MTS also unveiled an upgrade to its Nano Bionix, which more accurately tests polymers, biomaterials, and nanomaterials for small defects that can create strain in a material.

REASONS TO BE BEARISH
► MTS faces competition from larger equipment companies such as Veeco and FEI, as well as from smaller companies like Hysitron.

WHAT TO WATCH FOR   If the company can continue to line up some of the major manufacturers who are now working with nanomaterials—such as the major automobile manufacturers—to use its equipment, investors may want to consider adding shares.

CONCLUSION   Outlook is bullish. MTS is a solid, profitable company and is a low-risk investment. Investors wanting more exposure to nanotechnology, however, should stay with FEI or Veeco.

Publicly Traded "Pick 'n Shovel" Nanotechnology Companies

OBDCF	COMPANY	Obducat AM
	SYMBOL	OBDCF
	TRADING MARKET	Nordic Growth Market (but available throughout most major U.S. brokerages)
	ADDRESS	Geijersgatan 2A P.O. Box 580 Malmo, Sweden SE-201 25
	PHONE	46 40 3621 60
	CEO	Partrik Lundstrom
	WEB	*www.obducat.com*

DESCRIPTION   Obducat is an international leader in the emerging field of nanoimprint lithography (NIL), an emerging type of lithography that uses either a mold or a mechanical force to create a circuit. (Many industry observers feel such a technology may be the only method to design circuits at the 45, 32, 22, and sub-10 nanometer range.)

REASONS TO BE BULLISH
► According to the International Technology Roadmap for Semiconductors, the future production of electronic devices will require new lithography solutions, and Obducat is an early leader in nanoimprint lithography—having already sold dozens of its NIL presses to companies such as General Electric and major academic institutions.
► In 2004, the company updated its NIL System with new, improved features that will allow customers to perform dual functions simultaneously.
► Obducat is also part of a thirty member European consortium that will invest 16 million Euros to research and develop advanced production techniques for manufacturing nanostructures and multifunctional polymers. Any results derived from this consortium will remain the property of Obducat.

REASONS TO BE BEARISH
► To date, Obducat has not experienced an increase in sales, and its small size and relative dirth of capital leave it at the mercy of large competitors with deeper pockets.

continued

Obducat AM continued

► Nanoimprint lithography is expected to take hold when chip manufacturers begin operating in the 65 nm range. It is quite possible, however, that competing technologies such as FEI's dual beam system will slow the industry's move to nanoimprint lithography.

► Obducat currently lacks the human and capital resources to cover its customers on a 24-7 basis, and it faces stiff competition from start-ups Molecular Imprints and NanoNex.

WHAT TO WATCH FOR   If Obducat sells multiple pieces of its NIL equipment to a large chip manufacturer such as Intel or IBM, it will be a very bullish sign. On the other hand, if Molecular Imprints continues to sell its products to other large industrial customers (it has already sold units to KLA-Tencor and Motorola), it is not a positive sign for Obducat.

CONCLUSION   Outlook is bullish. Obducat is a small company playing in a field that has the potential to grow quite large. It is also selling scanning electron microscopes (about 40 percent of its business), and its customers include IBM, Siemens, and the U.S. Army. Its modest stock price, coupled with the fact that it is currently a leader in the field of nanoimprint lithography, offers a good risk-to-reward ratio for more aggressive investors.

Publicly Traded "Pick 'n Shovel" Nanotechnology Companies

POL	COMPANY	Polaron PLC
	SYMBOL	POL
	TRADING MARKET	London Stock Exchange
	ADDRESS	26 Greenhill Crescent
		Watford Business Park Watford, Hertfordshire, UK WD18 8X6
	PHONE	44.0.1923.495.495
	CEO	Joseph Stelzer
	WEB	*www.polaron.co.uk*

DESCRIPTION   Polaron is a holding company with four technology business components of which only one—Oxford Nanoscience LTD (which manufacturers three-dimensional atomic probes)—is related to nanotechnology.

REASONS TO BE BULLISH
- ► In the past year, Oxford Nanoscience has sold its atomic probes to a number of organizations in Japan, including the National Institute of Material Science.
- ► Its nanotechnology division only makes up approximately 15 percent of Polaron's revenues, but the company is profitable.

REASONS TO BE BEARISH
- ► Veeco, FEI, and Imago are all competitors, and it will be tough for Oxford, as a relatively small player, to distinguish itself in this field.

WHAT TO WATCH FOR   If U.S. and European companies working in the area of materials sciences begin purchasing Oxford's equipment, investors may want to consider investing in Polaron. On the other hand, if Imago's LEAP atomic probe appears to be gaining market share, it would be a bearish sign.

CONCLUSION   Outlook is bearish. Oxford Nanoscience's acceptance into the Japanese market is very promising, but until the company can line up additional business customers and until it becomes a more sizeable portion of Polaron's overall business, investors should treat the stock cautiously.

Publicly Traded "Pick 'n Shovel" Nanotechnology Companies

SMMX	COMPANY	Symyx Technologies
	SYMBOL	SMMX
	TRADING MARKET	NASDAQ
	ADDRESS	3100 Central Expressway Santa Clara, CA 95051
	PHONE	408-764-2000
	CEO	Steven D. Goldby
	WEB	*www.symyx.com*

DESCRIPTION   Symyx Technologies is a leader and pioneer in the field of high-throughput experimentation for the discovery of new materials for the chemical, electronics, life sciences, consumer goods, and automotive markets. In essence, the company has created a sophisticated and fast large-scale trial-and-error method for screening and developing new materials.

REASONS TO BE BULLISH

► By varying the chemical, physical, mechanical, electronic, optical, or catalytic properties of a material, manufacturers can gain new competitive advantages, and Symyx's software places the company at the forefront of an area that is likely to continue to grow.

► In 2004, the company's revenues increased nearly 30 percent, and profits rose from $2 million to nearly $8 million. The positive trend continued in the first half of 2005.

► Symyx already has licensing agreements with ExxonMobil, Dow Chemical, BP, Celanese, Merck, Pfizer, Honda, Eli Lilly, Agfa, and Unilever and UOP LLC. (Its agreement with ExxonMobil alone has already lead to $200 million in revenue, and its relationship with Dow could be just as lucrative.)

► The company has a strong intellectual property portfolio with over 160 U.S. patents and more than 400 worldwide patents.

REASONS TO BE BEARISH

► Symyx has not developed a specialty in the life sciences—arguably the largest growth sector—and is behind Avantium Technologies is this field.

continued

Symyx Technologies continued

► Smaller start-ups like Intematix (see page 66) are also developing sophisticated software to develop new materials for the electronics markets and could erode Symyx's market share.

WHAT TO WATCH FOR   The best sign of Symyx's strength will be the announcement of new partnerships. Another very positive development would be a statement that it has helped develop a new material for the storage of hydrogen. This would make fuel cells far more viable and if Symyx were to get a portion of the royalties on the material; it could lead to a very lucrative royalty stream.

CONCLUSION   Outlook is bullish. Traditional materials discovery relies on an expensive and time-consuming process of trial-and-error. Symyx's instrumentation, software, and methods have proven that they can help eliminate this problem by delivering faster results at a fraction of the cost. The bottom line is that nanotechnology is going to radically transform the materials sciences industry, and Symyx is well positioned to help lead the revolution.

TGAL	COMPANY	Tegal Corporation
	SYMBOL	TGAL
	TRADING MARKET	NASDAQ
	ADDRESS	2201 South McDowell Boulevard Petaluma, CA 94954-6020
	PHONE	707-763-5600
	CEO	Michael Parodi
	WEB	*www.tegal.com*

DESCRIPTION   Tegal designs, manufactures, markets, and services plasma etch and deposition systems that enable the production of advanced integrated circuits and memory devices. Tegal is considered a nanotechnology company because in late 2003 it purchased Simplus Systems, the developer of Nano Layer Deposition (NDL) technology, which is expected to help semiconductor companies make smaller transistors. Tegal is also listed on Merrill Lynch's Nanotech Index.

REASONS TO BE BULLISH
- Tegal has partnered with Sharp Laboratories in a joint development project designed to accelerate the adoption of integrated of next-generation solutions based on Simplus's NDL technology, and it has been awarded key patents for its NDL technology.
- The market for such next-generation equipment is expected to grow significantly. NDL might also have applications beyond the semiconductor industry to the telecommunications, data storage, and possibly even related life science industries.

REASONS TO BE BEARISH
- The company lost close to $5 million in FY2005, and no customers for its NDL have yet been announced.
- Genus and Applied Materials are working on similar technology.

WHAT TO WATCH FOR   Commercial sales of its NDL technology.

CONCLUSION   Outlook is bearish. Until customers for Tegal's NDL equipment are announced and the company begins to make significant inroads into high-volume markets, investors should tread cautiously.

Publicly Traded "Pick 'n Shovel" Nanotechnology Companies

UTEK	COMPANY	Ultratech Inc.
	SYMBOL	UTEK
	TRADING MARKET	NASDAQ
	ADDRESS	3050 Zanker Road San Jose, CA 95134
	PHONE	800-222-1213
	CEO	Arthur W. Zafiropoulo
	WEB	*www.ultratech.com*

DESCRIPTION   Ultratech (formerly known as Ultra Stepper) designs, manufactures, and markets photolithography equipment used in the fabrication of semiconductor and "nanotechnology" devices. Ultratech does not meet the official definition of nanotechnology, but because it purports to be a "nanotechnology company" and because it is listed on the Merrill Lynch Nanotech Index, a short review is in order.

REASONS TO BE BULLISH

▶ The company was profitable in 2004, its revenues increased by 10 percent for the year, and it has a substantial amount of cash on hand.

▶ Ultratech is well-diversified and provides tools for a variety of markets, including flat panel displays, thin film heads, optical components, sensors, inkjet print heads, and radio frequency identification devices, and the broader MEMS industry.

REASON TO BE BEARISH

▶ It faces stiff competition from competitors like Tegal, as well a number of smaller companies who are striving to develop the equipment necessary to fabricate the next generation of smaller integrated circuits.

▶ In the second quarter of 2005, the company announced reduced earnings due to delayed shipments.

WHAT TO WATCH FOR   If it announces a new line of equipment that can operate at the sub-65 nanometer and 45-nanometer range, investors may want to reconsider.

CONCLUSION   Outlook is bearish. With other companies such as Veeco, FEI, and Obducat more closely involved in nanotechnology, investors would be better off investing a portion of their portfolio in those companies. Longer term, Ultratech also risks losing market share to more aggressive start-ups like Molecular Imprints.

Publicly Traded "Pick 'n Shovel" Nanotechnology Companies

VARI		
	COMPANY	Varian, Inc.
	SYMBOL	VARI
	TRADING MARKET	NASDAQ
	ADDRESS	3120 Hansen Way Palo Alto, CA 94304
	PHONE	650-213-8000
	CEO	Garry Rogerson
	WEB	*www.varianinc.com*

DESCRIPTION   Varian develops, manufactures, and services a variety of scientific instruments, including mass spectroscopes and nuclear magnetic resonance (NMR) equipment, for the electronics and life sciences sectors. The company is one of only eleven companies in the Activest Lux Nanotech Mutual Fund.

REASONS TO BE BULLISH
► Varian has experienced solid growth for the past five years. In 2004, its sales grew 8 percent, and it had posted a profit of $58 million. The positive trend continued in the first half of 2005.
► In late 2004, it also purchased Magnex Scientific and improved its NMR equipment.

REASON TO BE BEARISH
► Varian faces a good deal of competition from Agilent and PerkinsElmer.

WHAT TO WATCH FOR   NMR can play a very important role in the fields of proteomics and genomics by identifying the 3-D structure, function, and dynamics of proteins, nucleic acids, and other biological molecules. As such, it has an important role to play in drug discovery. Investors will want to keep an eye on whether major pharmaceutical manufacturers are purchasing Varian's equipment for this purpose.

CONCLUSION   Outlook is bullish. In the short term, Varian is likely to experience positive but modest growth. Longer term, it will be incumbent on the company to conduct the R & D necessary to stay competitive in the rapidly changing electronic and life sciences sectors.

VECO	COMPANY	Veeco
	SYMBOL	VECO
	TRADING MARKET	NASDAQ
	ADDRESS	100 Sunnyside Boulevard Woodbury, NY 11797-2902
	PHONE	516-677-0200
	CEO	Edward A. Braun
	WEB	*www.veeco.com*

DESCRIPTION   Veeco is a leading provider of metrology equipment used in the data storage, semiconductor, and telecom/wireless industries. In 2005, approximately 34 percent of its revenue came from the data storage market; 27 percent from sales to research institutions; 24 percent percent from the wireless industry; and 15 percent from the semiconductor industry.

REASONS TO BE BULLISH

▶ Veeco controls approximately 70 percent of the world market in high-end metrology systems—primarily Atomic Force Microscopes (AFM)—which are used by chipmakers to locate tiny defects and measure materials at the nanoscale, and its equipment is installed at every one of the world's leading chipmakers, including Intel, IBM, and Motorola.

▶ The company is well positioned for future growth because the leading chipmakers need to manufacture next-generation devices at the 65 nm and then 45 nm levels, and they will need to continue to rely on Veeco's equipment.

▶ The company has recently developed two centers in China, which stands to be a big player in the development of nanotechnology.

▶ Veeco received a $6.6 million grant from the U.S. government to partner with Dow Chemical to develop a nanomechanical measurement instrument that can potentially help in the development and creation of new materials. If successful, the product would allow material scientists to rapidly develop and commercialize new materials with breakthrough properties.

▶ It has over $100 million cash-on-hand for the purpose of strategic acquisitions.

Veeco continued

REASONS TO BE BEARISH

► A sizeable portion of its business is still tied to the two very cyclical industries: semiconductor and telecom/wireless. To the extent that those markets slow, Veeco will be adversely affected. In the first quarter of 2005, orders were down from $117 million to $98 million.

► Veeco's other market—the hard disk drive—is essentially a commodity market, and profits are marginal.

► The company has pursued an extensive acquisition strategy and may have a difficult time integrating its acquisitions.

WHAT TO WATCH FOR   If Veeco can develop an entry-level AFM, its products will become accessible to a host of new customers. Secondly, if RFIDs (radio frequency identification tags) successfully take root in the commercial marketplace, Veeco, as a leading equipment supplier to the main manufacturers of RFIDs, will stand to gain.

CONCLUSION   Outlook is bullish. Veeco is a solid long-term investment. In 2006, an estimated $10 billion will be spent on nanotechnology research development. Much of this money will be spent on equipment, and Veeco, as world leader, will benefit. Moreover, because its equipment is used in diverse fields—life sciences, materials sciences, and semiconductor industries—it can better handle the cyclical swings of the semiconductor industry.

At the present time, none of the following companies are publicly traded. Investors, however, need to beware of them for three reasons. First, some may go public and might represent a good investment. Second, they may become acquisition targets, and the company that purchases them may gain a competitive advantage that would be reflected in its stock price. Third, their technology is "disruptive" and could take away market share from an existing company and cause a concomitant decrease in its stock price.

COMPANY	Arryx, Inc.
INVESTORS	Draper Fisher Jurvetson, ARCH Development Corp., Haemonetics Corp.
ADDRESS	316 North Michigan Avenue, Suite CL20 Chicago, IL 60601
PHONE	312-726-6675
CEO	Lewis Gruber
WEB	*www.arryx.com*

DESCRIPTION   Arryx develops and manufactures holographic optical trapping (HOT) technology that capitalizes on the ability of optical traps to manipulate matter. The technology can also be used to measure tiny forces of biological interest. The benefit of HOT is that it can create an enormous number of traps and, as such, represents a quantum improvement over standard optical trapping because it can do it much faster and in three dimensions.

WHY IT IS DISRUPTIVE   Think of Arryx's HOT technology as "laser tweezers" capable of moving and manipulating cells, organelles, and molecules. Its BioRyx 200, which is already on the market, can isolate cells and molecules, detect and measure the presence of materials, and possibly even help make sensors to detect biological and chemical agents. The technology has a host of potential applications in diagnostics, drug discovery, and material testing markets. The HOT technology may also be able to build a variety of future products including telecommunication components and sophisticated sensors.

WHAT TO WATCH FOR   New partnerships will be integral to Arryx's success. If it can continue to line up new partners and should the company go public, it would be worth considering as a possible investment for those with a high tolerance for risk.

CONCLUSION   Arryx is definitely a company to watch. In late 2004, it secured a $5 million equity investment from Haemonetics (a maker of automated blood processing systems) and, in return, received an exclusive worldwide license to use Arryx's technologies for use in blood processing. It validates Arryx's technology and could result in modest royalty payments of $12.5 million over the next five years. Arryx currently has no direct competition, although other companies such as Zyvex (see page 241) are working to create alternative nanoscale manipulation systems.

COMPANY	Avantium Technologies BV
INVESTORS	Shell Chemicals, Eastman Chemicals, GlaxoSmithKline, Pfizer, and others
ADDRESS	Zekergstraat 29, 1014 BV P.O. Box 2915, 1000 CX Amsterdam, Netherlands
PHONE	31-20-586-8080
CEO	Edwin Moses
WEB	*www.avantium.com*

DESCRIPTION Avantium is a research and development company specializing in the creation of new materials and catalysts for the pharmaceutical, biotech, and chemical industries.

WHY IT IS DISRUPTIVE Avantium uses high throughput experimentation, analysis, and simulation technology to do large-scale trial-and-error experimentation on new materials. Instead of doing one experiment at a time, the company can do thousands. It also claims to have the world's largest commercially available library of materials and catalysts in the world.

WHAT TO WATCH FOR Symyx (see pages 54–55) is a competitor of Avantium's and bears watching. If it can gain entry to the pharmaceutical sector, Avantium could suffer.

CONCLUSION At the present time, Avantium has a much stronger focus on the pharmaceutical side and given its current work with Pfizer, GlaxoSmithKline, and Astra-Zeneca it is unlikely to lose this competitive advantage anytime soon. In the event it should go public, an investment would be warranted. Investors are also encouraged to keep an eye on those companies who are partnering with Avantium because it could give them a leg up on their competitors.

COMPANY	Hysitron
INVESTORS	Private
ADDRESS	10025 Valley View Road Minneapolis, MN 55344
PHONE	952-835-6366
CEO	Thomas Wyrobek
WEB	*www.hysitron.com*

DESCRIPTION   Hysitron is a world leader in designing, producing, and servicing nanomechanical test instruments for advanced research and industrial applications.

WHY IT IS DISRUPTIVE   Hysitron's TriboIndentor—which makes a small indentation in the surface of a material to measure properties such as wear, thickness, hardness, adhesive strength, and elasticity—will become increasing useful (and necessary) as companies continue to build devices, surface coatings, and thin films in the nanometer range. Hysitron's equipment offers a way to ensure uniformity, test for quality, and, ultimately, improve performance.

WHAT TO WATCH FOR   Hysitron sells more nanoindentation equipment than all of its competitors combined, but it still faces tough competition from MTS Systems and others.

CONCLUSION   Revenues have grown from $6 million in 2000 to $14 million in 2004 and an estimated $25 million for 2005, and its upward path should continue as more research labs and companies demand its products.

COMPANY	Imago Scientific Instruments
INVESTORS	Draper Fisher Jurvetson, Infineon Ventures
ADDRESS	6300 Enterprise Lane Madison, WI 53719
PHONE	608-274-6880
CEO	Timothy J. Stultz
WEB	*www.imago.com*

DESCRIPTION   Imago manufactures an advanced microscope called the Local Electrode Atom Probe (LEAP) that, among other things, is being used to create "high security" steel that might be able to resist possible terrorist attacks by helping to rapidly analyze at the atomic level the precise chemical composition of the material.

WHY IT IS DISRUPTIVE   The LEAP can achieve analysis rates 720 times faster than traditional SEM and TEM, which allows chip manufacturers to analyze and detect defects in their chip much sooner than is currently possible. Some estimates suggest that this feature could save a large chip manufacturer $100 million annually. More importantly, perhaps, the LEAP can be used by material scientists to gain a unique three-dimensional and atomic level view of materials—making it a potentially powerful drug discovery tool.

WHAT TO WATCH FOR   If life science companies begin to purchase the LEAP to analyze how biological materials are working, it should be considered a very bullish sign. Investors need to keep this in mind because if a large company acquires Imago it could give that company a strong presence in the life science sector—a high-margin and growing area.

CONCLUSION   Imago is a very promising company. The company has ongoing relations with some of the leading U.S. government laboratories, and in September 2004, Seagate announced it had adopted Imago's LEAP technology. It also holds some valuable intellectual property and appears well positioned for future growth, although competitors in the SEM and TEM industries are unlikely to allow Imago to go unchallenged. If it goes public, investors should seriously consider investing in the company. A more likely scenario, however, is that Veeco, FEI, or Hitachi or another large equipment supplier will acquire the company.

COMPANY	Intematix Corp.
INVESTORS	Draper Fisher Jurvetson
ADDRESS	351 Rheem Boulevard Moraga, CA 94556
PHONE	925-631-9005
CEO	Ruediger Stroh
WEB	*www.intematix.com*

DESCRIPTION   Intematix develops and manufactures high-throughput screening technology specializing in the creation and characterization of new electronic materials.

WHY IT IS DISRUPTIVE   The research and development of component materials for the electronics industry is one of the most common bottlenecks in new product development. Intematix's proprietary process for creating new electronic materials is reportedly 50 to 100 times faster than conventional methods and suggests that Symyx's technology may be challenged.

WHAT TO WATCH FOR   Intematix doesn't yet have any corporate partners. If it announces an agreement with an established industrial partner for high-volume manufacturing, it will be a significant milestone.

CONCLUSION   The business model it is seeking to follow has already been successfully demonstrated by Symyx (which does it for chemical compounds), and Intematix's process suggests it has applications in the telecommunication, automotive, aerospace, and material science sectors. If the company can demonstrate that once it has discovered a new material, it can license those materials or, alternatively, partner with a large industrial company to manufacture the material, Intematix could either go public or become an acquisition candidate. In either case, it bears watching. The company will face competition from Symyx.

COMPANY	JPK Instruments
INVESTORS	IBB Beteiligungsgesellschaft
ADDRESS	Bouchestrasse 12 Berlin, Germany 12435
PHONE	49 30 5331 12542
CEO	Frank Pelzer
WEB	*www.jpk.com*

DESCRIPTION   JPK has designed and manufactured an Atomic Force Microscope—named the Nano-Wizard.

WHY IT IS DISRUPTIVE   The Nano-Wizard combines the technology of an AFM with visible light microscopy. This combination allows researchers to view and examine living samples such as DNA, proteins, and other biological samples. The technology has been designed specifically for the life science sector.

WHAT TO WATCH FOR   JPK currently has no presence in the United States. If it can address that shortcoming or if major pharmaceutical firms were to purchase JPK's equipment, it will be a bullish signal. The company's growing global presence makes JPK an attractive acquisition target for a larger equipment supplier.

CONCLUSION   The life science sector needs equipment like the Nano-Wizard, and the market for such equipment is likely to grow considerably in the years ahead as a result of increased investment in cellular research and drug discovery. The fact that JPK now has a growing distribution network with agreements in place in Great Britain, China, Taiwan, South Korea, Japan, and Canada suggests it is finding growing markets for its products.

COMPANY	Molecular Imprints
INVESTORS	KLA-Tencor, Motorola, Lux Capital, Draper Fisher Jurvetson, Harris & Harris, Brewer Science, Huntington Ventures, Alloy Ventures, Lam Research
ADDRESS	1807-C West Braker Lane Austin, TX 78758-3605
PHONE	512-339-7760
CEO	Dr. Norman E. Schumaker
WEB	*www.molecularimprints.com*

DESCRIPTION   Molecular Imprints is the world's leading manufacturer of step-and-flash imprint lithography (S-FIL)—a stamping lithography technique that is capable of delivering resolution to 20 nm and below at a high speed and low cost (estimated to be ten times less expensive than today's state-of-the-art). In 2005, the *EETimes* awarded the company its "most promising new technology" award.

WHY IT IS DISRUPTIVE   Semiconductors will get increasingly crowded as transistors continue to get smaller. Molecular Imprints has the ability to help companies make transistors as small as the 20 nm level. Furthermore, because its S-FIL technology can operate at lower temperatures and pressures than that of its leading competitors, it is well positioned to deliver a low-cost, low-complexity alternative to today's optical lithography tools.

WHAT TO WATCH FOR   It is possible that radical improvements in extreme ultraviolet lithography (EUV) may be able to supersede NIL technology or that competitors like Obducat, NanoNex, and the EV Group may develop superior technology.

CONCLUSION   Molecular Imprints has an extremely strong intellectual property portfolio with close to 100 patents and has a $36 million grant from the National Institute of Standards and Technology to develop nanoimprint lithography infrastructure for low-cost, high-throughput replication at the 65 nm node and below. More important, it has already lined up some key partners and customers, including KLA-Tencor and Motorola. If the company goes public in 2006, investors should strongly consider an investment. In the interim, the publicly traded venture capital firm Harris & Harris (NASDAQ: TINY) (see pages 175–76) offers an indirect way for the individual investor to buy at least a portion of this company.

COMPANY	NanoInk, Inc.
INVESTORS	Galway Partners LLC, Lurie Investment Fund LLC
ADDRESS	1335 Randolph Street Chicago, IL 60607
PHONE	312-525-2900
CEO	Cedric Loiret-Bernal
WEB	*www.nanoink.net*

DESCRIPTION NanoInk is the designer, developer, and manufacturer of Dip-Pen Nanolithography (DPN), a unique nanofabrication technique that uses an AFM to write nanoscale patterns on substrates using the atoms and molecules as the "ink."

WHY IT IS DISRUPTIVE DPN, because it can place thousands and, longer term, maybe even millions or more of atoms and molecules exactly where they are needed, has the potential to usher in an era of "bottom-up" manufacturing—that is, build small devices with atomic precision. In the short term, DPN could potentially revolutionize medical research and diagnostics by allowing for the creation of small, fast, ultrasensitive point-of-care diagnostics.

WHAT TO WATCH FOR NanoInk will need to demonstrate that its equipment has large scale, parallel patterning capability that many industrial uses will require. If it does, look for future corporate partners to emerge. Investors will then want to look for licensing/royalty agreements. In the near term, investors should watch if anything tangible develops out of the company's grant from the National Institutes of Health to create ultrasmall DNA arrays.

CONCLUSION NanoInk holds key patents, including one in the area of precision control at the molecular level, which will allow the creation of devices from the "bottom-up." Moreover, because its DPN equipment is high resolution, low cost, relatively easy to use, and can employ a variety of materials, it has a lot of practical, near-term applications. For instance, NanoInk has partnered with Carbon Nanotechnologies in a joint development agreement to employ single walled carbon nanotubes in next-generation, nanofabricated devices. If the company decides to go public, investors should consider it not only for its near-term potential in the DNA array market, but also for its long-term potential to revolutionize how small devices are manufactured.

COMPANY	NanoNex
INVESTORS	Hitachi
ADDRESS	1 Deer Park Drive, Suite O Monmouth Junction, NJ 08852
PHONE	732-355-1600
CEO	Stephen Chou
WEB	*www.nanonex.com*

DESCRIPTION   NanoNex develops and manufacturers nanoimprint lithography solutions including tools, masks, polymers, and processes.

WHY IT IS DISRUPTIVE   NanoNex's equipment costs significantly less than today's step-and-scan systems, and that makes it an attractive option for today's semiconductor companies who will eventually need to create sub-10 nanometer features and develop three-dimensional patterns. Longer term, it is possible that its equipment can be used by biotech researchers. It is also feasible that NanoNex's devices can be used to create next-generation optical and data storage devices.

WHAT TO WATCH FOR   Obducat is partnering with GE, Molecular Imprints with Hewlett-Packard and Motorola, and NanoNex with Hitachi. If any of these companies switch partners, it could serve as a validation/repudiation of its technology. Also, if biotechnology companies begin to demonstrate an interest in NanoNex's equipment, it would be a bullish sign.

CONCLUSION   Molecular Imprints's competing technology can reportedly operate at lower temperatures and pressures and is it also believed to be faster. The fact that NanoNex has entered into a partnership with Hitachi to sell its equipment is a positive development; however, it will need to find customers. Until it does, investors interested in nanoimprint lithography are encouraged to look at Obducat and Molecular Imprints (if it goes public).

COMPANY	nPoint
INVESTORS	Angel Funding
ADDRESS	1617 Sherman Avenue Madison, WI 53704
PHONE	608-310-8770
CEO	John Biondi
WEB	*www.npoint.com*

DESCRIPTION   nPoint designs and manufactures nanopositioning components and systems used in atomic force microscopes (AFMs) and scanning electron microscopes (SEMs).

WHY IT IS DISRUPTIVE   nPoint's add-on components are capable of rapid, precise, and repeatable positioning and motion at the nanometer scale—a feature that improves significantly the utility of AFMs and SEMs.

WHAT TO WATCH FOR   Veeco and FEI could very well decide to develop their own specialty in this area, and with their deep pockets, they could make it difficult for nPoint to compete. If, however, nPoint can develop a retrofit component that works with existing AFMs, its client base could grow significantly.

CONCLUSION   nPoint bears watching. The company is already supplying components to the leading AFM suppliers. The type of tools it provides currently represent only 10 percent of the cost of an AFM, but as the components become increasingly more sophisticated, that figure will grow—some estimates suggest it may approach 50 percent annual growth. If its equipment continues to be demanded by leading researchers and companies in the nanotechnology and biotechnology industries, nPoint could become an attractive acquisition target for Veeco or FEI.

Summary

The amount of money being invested in nanotechnology by governments, universities, corporations, and private start-ups is expected to exceed $10 billion in 2006 and will only grow in the years ahead as more fields move into the nano realm. Equipment suppliers to the nanotech field are nicely positioned to grow and prosper from this increased activity.

Investor enthusiasm should be tempered by two factors. First, while equipment and tools are vital for the field's success, investors should expect—with a few exceptions—to receive solid but not extraordinary returns on their investments. Second, as should have been clear from reading the profiles, many of the companies are developing similar technologies, and the marketplace will likely only be able to accommodate a few winners in each field.

A related point is that although many of the companies have very advanced technologies that does not mean they cannot also be disrupted. For instance, a piece of equipment that is necessary to build chips at the 65 nm range may be obsolete at the 45 nm range. The time frame between these two events will be less than two years.

Remember that while the future may belong to the "small" in the nanotechnology era, it will still be vulnerable to the fast.

"GE is investing in the future of materials. We are investing a lot of our efforts in nanotechnology. There are some non-believers, but I think we will see some real applications in the next 3–5 years and some real traction in the next decade. By 2020, this can be a $35 billion industry . . . Nanotechnology will have a dominant impact."

—Jeffrey Immelt, CEO, General Electric

Chapter 4

A Precious Commodity: Nanomaterials

Most people are familiar with the famous scene in the 1965 hit movie *The Graduate* when Dustin Hoffman was pulled aside at his graduation party and offered one word of advice: "Plastics." If the scene were replayed today, the word "nanotechnology" could easily be inserted in lieu of plastics. Nanotechnology will revolutionize tomorrow's products in the next half century in much the same way plastics did in the last half century.

The analogy is apt for another reason. That is because while polymers have found their way into a host of products—everything from soft drink bottles and diapers to automobiles components and implantable medical devices—the price of most plastic resins has dropped to commodity levels as the technological know-how and manufacturing proficiency has increased. The future for nanoparticles, nanoclays, carbon nanotubes, dendrimers, quantum dots, and a variety of other nanomaterials could very well be much the same. In fact, it is already happening. The price of a number of specialty nanomaterials has steadily decreased over the past few years. Certain carbon nanotubes that were once $1,000 per gram are now going for less than $100—and their price continues to drop.

This is a positive development in the sense that the reduction in price will allow the materials—which are one hundred times stronger than steel and just one-sixth the weight—to be incorporated into an increasing number of products. It is a negative development in that the profit margins for the suppliers of those materials will be squeezed ever tighter.

This development suggests that unless companies can produce their nanomaterials in bulk quantities at competitive prices or, alternatively, partner with—or license their technology to—a larger company to produce their nanomaterials, they will be marginalized by the few companies who can achieve the necessary economies of scale.

Not all nanomaterials producers, however, are destined to play either second fiddle to bigger companies or file for bankruptcy. The value of their respective nanomaterials lies in their inherent ability to increase the value of the product in which they are being used. If the nanomaterial companies can find a niche in a large enough market—for instance, automobile components, fuel cell technology, or medical diagnostics—they can become very profitable. Similarly, those companies who are producing at commodity price levels can be profitable if their volumes are large enough.

The majority of this chapter is dedicated to companies that have the potential of achieving one or the other. In general, however, investors will want to tread cautiously in the area of nanomaterials because of the risk of commoditization and because there are so many companies competing for the same end user. For instance, Nanophase, Oxonica, and Five Star Technologies are all producing nanoparticles for sun screen lotions. Catalytic Solutions and Nanostellar are both developing nanomaterials to replace platinum in catalytic converters, while Evident and Quantum Dot are both manufacturing nanocrystals for medical diagnostics. Meanwhile, Carbon Nanotechnologies and SouthWest Nanotechnologies are both producing carbon nanotubes for flat panel displays, while Eikos and Hybrid Plastics are seeking to apply carbon nanotubes for space-based applications.

All of these companies are necessary for the investor to be aware of because their nanomaterials could give the companies *using* their materials a decided advantage in the marketplace—an advantage that could well be reflected in those companies' stock prices.

This chapter begins with profiles of the three publicly traded small cap nanomaterials companies (Chapter 5 has profiles on some of the other large suppliers, including BASF, Degussa, and Frontier Carbon) but will devote most of its attention to those privately held companies who are helping to deliver the first wave of nanotechnology-enabled products to the market today.

AMR.TO		
	COMPANY	AMR Technologies
	SYMBOL	AMR.TO
	TRADING MARKET	Toronto Stock Exchange
	ADDRESS	121 King Street West, Suite 1740 Toronto, Ontario Canada, M5H 3T9
	PHONE	416-367-8588
	CEO	Peter Gundy
	WEB	*www.amr-ltd.com*

DESCRIPTION   AMR produces, processes, and develops rare earth and zirconium-based materials, nanoparticles, and nanocoatings for applications in televisions, cell phones, and CD/DVD players, as well as fuel cells and catalytic converters.

REASONS TO BE BULLISH

► AMR has demonstrated consistent growth and, when compared with the market valuations of some of its U.S. competitors, offers an attractive valuation.

REASONS TO BE BEARISH

► Confidentiality agreements keep the company from discussing either to whom it is supplying many of its products or for what purpose. This makes the stock difficult to evaluate.
► It also faces tough competition in every one of its announced markets.

WHAT TO WATCH FOR   In 2004, the company announced that it had developed a novel form of cerium oxide nanoparticle and claimed this would open some lucrative markets. If these are used by a major manufacturer for a large market—such as flat panel displays, fuel cells, or catalytic converters—it will be a bullish sign and would warrant reconsideration as an investment.

CONCLUSION   Outlook is bearish. AMR is a good solid company and represents a better investment than some of the other publicly traded nanomaterials companies, but until the company can be more forthcoming about its customers and their products, investors should treat the stock cautiously.

ACO	COMPANY	Nanocor, Inc. (a wholly owned operating subsidiary of Amcol International Corporation)
	SYMBOL	ACO
	TRADING MARKET	NYSE
	ADDRESS	1500 West Shure Drive Arlington Heights, IL 60004
	PHONE	847-394-8730 (Amcol Headquarters)
	CEO	Peter Maul
	WEB	*www.nanocor.com*

DESCRIPTION   Nanocor is the largest global manufacturer and supplier of nano-clays—an additive that can be incorporated into resins to increase the strength, heat resistance, and gas barrier properties of a variety of plastics.

REASONS TO BE BULLISH
- In 2004, Amcol's sales rose 23 percent, and its profits increased to $28 million. The positive trend continued in the first part of 2005.
- Nanocor's nanoclays are already being used in package containers, appliance casings, automobile components, and various consumer products.
- It has successfully licensed its technology to Honeywell Plastics, Bayer AG, and Eastman and has a strong intellectual property portfolio.

REASONS TO BE BEARISH
- Nanocor is currently a small—albeit growing—part of Amcol's overall business.
- Plastic resins are a commodity business, and Nanocor may find it difficult to generate market-beating returns against competitors such as Southern Clay and Hybrid Plastics.

WHAT TO WATCH FOR   If Amcol spins off Nanocor (which it has considered in the past), the new entity would be worth considering as an investment.

CONCLUSION   Outlook is neutral. Amcol is a solid company, and the potential market for nanoclays is significant. However, until Nanocor either separates from Amcol or becomes a more significant part of Amcol's overall business, investors should recognize that they would only be making a small investment in a nanomaterials company.

Publicly Traded Nanomaterials Companies

NANX	COMPANY	Nanophase Technologies Corporation
	SYMBOL	NANX
	TRADING MARKET	NASDAQ
	ADDRESS	1319 Marquette Drive Romeoville, IL 60446
	PHONE	630-771-6700
	CEO	Joseph Cross
	WEB	*www.nanophase.com*

DESCRIPTION In 1997, Nanophase was the first nanomaterials company to go public and is a leading developer and manufacturer of nanomaterials.

REASONS TO BE BULLISH

► The overall market for nanomaterials is expected to grow, and the market for specialty nanomaterials—in which Nanophase specializes—is less subject to the cyclical downturns in the economy.

► The potential market for Nanophase's nanomaterials is large and includes personal care, sunscreens, abrasion-resistant applications, environmental catalysts, antimicrobial products, and semiconductor polishing.

► Nanophase has existing partnerships with BASF to create nanomaterials for sunscreens and personal care applications and Rohm & Haas to supply nanomaterials for the semiconductor polishing market.

► The company has a strong intellectual property portfolio with key patents in the areas of nanoparticle synthesis and surface treatment technologies, and its seasoned management team has continued to improve operations and focused on innovation.

REASONS TO BE BEARISH

► Nanophase has yet to be profitable in eight years, and it is currently trading near its original IPO level.

► It will likely find itself in competition against industrial giants like Degussa, as well as start-ups such as Oxonica and Five-Star Technologies.

continued

Nanophase Technologies Corporation continued

WHAT TO WATCH FOR If Nanophase is successful in diversifying and strengthening its number of strategic partners and can also move into new areas (e.g., antimicrobial products or environmental catalysts), it could become profitable and would become a more favorable investment.

CONCLUSION Outlook is bearish. Until Nanophase can demonstrate an ability to turn a profit, investors would be wise to treat the company with caution. Furthermore, given the number of start-up companies in the field who threaten to take away market share in key areas like nanomaterials for personal care, as well as larger companies like Degussa who can drive down the price on nanomaterials, its future prospects don't look promising. The one advantage Nanophase does have is that due to its relative longevity in the nanomaterials business it has a leg up on new nanomaterials companies who may underestimate the time and difficulty of scaling and producing unique nanomaterials.

Publicly Traded Nanomaterials Companies

OXN	COMPANY	Oxonica
	SYMBOL	OXN (London Stock Exchange)
	ADDRESS	Unit 7, Begbroke Science and Business Park Kidlington, United Kingdom OX5 1PF
	PHONE	00.44.1865.856.728
	CEO	Dr. Kenneth Matthews
	WEB	*www.oxonica.com*

DESCRIPTION   Oxonica is a leading producer of nanomaterials for catalysis, personal care, and biodiagnostics. Its subsidiary, Cerulean International, distributes the company's Envirox nanocatalyst that helps fuel burn cleaner and more efficiently.

WHY IT IS DISRUPTIVE   Oxonica's unique OPTISOL nanoparticles absorb ultraviolet radiation without forming cancer-causing free radicals. Such an advance could revolutionize the skin care industry by virtually eliminating the generation of free radicals. Similarly, Environ, which has been successfully tested and demonstrated to improve the fuel efficiency in diesel buses by 10 percent, has the potential to lead to significant increases in fuel efficiency and emission reductions. If the technology can be expanded to the broader automobile industry, it could lead to the type of high-volume production that could yield sizeable profits.

WHAT TO WATCH FOR   Oxonica will face competition from both midsize competitors as well as larger corporations such as Degussa. The effect could be that Oxonica's products face severe downward price pressure. If, however, it can develop biodiagnostic quantum dots—which it is working on and for which there are higher margins—it can lead to significant profits. But even in this area it must be noted that Oxonica would be competing with well-established companies like Quantum Dot Corp. (see page 99) and Evident Technologies (see page 87).

CONCLUSION   Outlook is bullish. Oxonica's products have been tested successfully in both Europe and Asia, and the company has a strong intellectual property portfolio. It also has a partnership with BASF that could lead to high-volume production.

COMPANY	Aspen Aerogels
INVESTORS	Reservior Capital Group LLC and Rockport Capital Partners LP
ADDRESS	30 Forbes Road Northborough, MA 01532
PHONE	508-691-1111
CEO	Don Young
WEB	*www.aerogel.com*

DESCRIPTION   Aspen Aerogels is a leader in the production of aerogel technology—nanoporous, lightweight materials that exhibit extraordinary low-thermal and acoustic conductivity.

WHY IT IS DISRUPTIVE   Aspen's aerogels (also referred to as "frozen smoke" because they are 95 percent air) have a variety of applications in consumer, commercial, and military markets. Because they have anywhere from two to ten times better thermal and acoustical insulating power than traditional insulators like foam and fiberglass, they can be used in everything from shoe inserts to materials that mask the heat of an engine. They can also be used to insulate homes and buildings as well as pipelines.

WHAT TO WATCH FOR   Although its insulating material is reportedly three times as effective as 3M's Thinsulate, it will face competition from 3M. It will also face stiff competition from Cabot (see pages 122–23), which also is a large manufacturer of aerogels.

CONCLUSION   Aspen is already working with NASA, the U.S. Army, DuPont, Boeing, and GM, and appears well positioned for future growth. If it can continue to line up additional corporate partners, it has a very promising future and investors are encouraged to give it serious consideration in the event it decides to go public.

COMPANY	Cap-XX, Inc.
INVESTORS	Intel Capital, ACER Technology Ventures, ABN AMRO Capital, Technology Venture Partners, Walden International
ADDRESS	12 Mars Road, Units 9 and 10 Lane Cove, NSW 2066 Australia
PHONE	61.2.9420.0690
CEO	Anthony Kongats
WEB	*www.cap-xx.com*

DESCRIPTION   Cap-XX is an Australian designer and manufacturer of carbon-based supercapacitors—which are power storage devices that bridge the gap between batteries and capacitors.

WHY IT IS DISRUPTIVE   Cap-XX's supercapacitors are manufactured with nanoscale materials, which can provide greater storage capacity and more powerful discharge capabilities. These attributes will allow cell phones, PDAs, and computer notebooks to use less energy, last longer, and be made thinner. The ability exists to further refine the nanomaterials to gain even greater storage and discharge capacities, suggesting that supercapacitors might find applications in the automotive market.

WHAT TO WATCH FOR   Additional partnerships will be a good indicator of its technological proficiency. If Cap-XX announces that its supercapacitors are being used in medical devices, digital cameras, or the automotive market, it would be a bullish sign.

CONCLUSION   Given the company's strong portfolio of intellectual property and the fact that it has already established a partnership with Intel Corp, Cap-XX has the resources, technical expertise, and relationships to become a market leader in the supercapacitor niche. It will face stiff competition from NEC and others, but investors should put it on their radar screen in the event the company goes public.

Privately Held Nanomaterial Companies

COMPANY	Carbon Nanotechnologies
INVESTORS	Richard Smalley, Gordon Cain, William McGinn, Bob Gower, other individual investors
ADDRESS	16200 Park Row Houston, TX 77084-5195
PHONE	281-492-5707
CEO	Bob Gower
WEB	*www.cnananotech.com*

DESCRIPTION   Carbon Nanotechnologies (CNI) was founded by Dr. Richard Smalley, the 1996 Nobel Prize winner in chemistry, and is among the world leaders in the development and production of high-quality carbon nanotubes—also known as buckytubes.

WHY IT IS DISRUPTIVE   Carbon nanotubes are imbued with a host of extraordinary properties. Their high strength-to-weight ratio, small size, and high conductivity open up a host of potential applications in flat panel displays, conductive polymers, high-strength materials, lithium ion batteries, solar energy converters, and electronics. To the extent that these materials change the television manufacturing, plastics, battery, energy, and electronics industries, a great many existing manufacturers will find they no longer have the materials or products to compete effectively in these rapidly changing markets.

WHAT TO WATCH FOR   In late 2004, CNI acquired C Sixty (a small company employing fullerenes for potential pharmaceutical applications), which may open up some opportunities in the life sciences sector. It has also received a grant from the Advanced Technology Program to develop fuel cell technology as well as a grant from NASA to help develop "Quantum Wires" with Rice University. Both bear watching because they open up CNI to some potentially lucrative revenue streams.

continued

Carbon Nanotechnologies continued

Investors also need to monitor the environmental issues surrounding carbon nanotubes. If they are found to be harmful to the environment of human health, the negative consequences are obvious. Also, potential investors need to be aware of legal issues surrounding who owns the patents on carbon nanotubes. In 2004, NEC—which first discovered carbon nanotubes in 1993—announced its intention to require the producer of any carbon nanotubes to seek a license through the company. If its claim is upheld, it could serve to dampen CNI's prospects for success.

CONCLUSION   Carbon Nanotechnologies currently has over 450 government, academic, and corporate customers and has licensed its intellectual property to Dupont for use in field-emission flat panel displays. It has also partnered with Sumitomo Corporation to sell its products into the Japanese and South Korean market and has partnered NanoInk (see page 69) to explore using its carbon nanotubes in NanoInk's Dip Pen Lithography to manufacture next-generation electronic devices. With its well-established customer base, strong partnerships, and deep intellectual property portfolio, CNI is well positioned to be one of the major players in the nanotechnology arena in the years ahead. If the company goes public, investors should consider a possible investment. The downside is that CNI faces stiff competition from NEC, Hitachi, and SouthWest Nanotechnologies, and it must confront the possibility that carbon nanotubes will face downward pricing pressure as production and competition increases.

COMPANY	Catalytic Solutions, Inc.
INVESTORS	Rockport Capital Partners, BASF Venture Capital, EnerTech, NGEN, Cycad Group, JP Morgan, Cinergy Ventures, GE Power Systems, SAM Group, Presidio Venture Partners
ADDRESS	1640 Fiske Place Oxnard, CA 93033
PHONE	805-486-4649
CEO	William Anderson
WEB	*www.catsolns.com*

DESCRIPTION   Catalytic Solutions (CSI) develops and produces nanostructured materials with superior catalytic performance—including for catalytic converters.

WHY IT IS DISRUPTIVE   The market for catalysts—especially the catalytic converter market—is very large. Moreover, many of today's catalytic converters use platinum, palladium, and rhodium, which are all very expensive. CSI's Mixed Phase Catalyst (MPC) technology not only has superior performance capabilities, but it also has significant cost advantages because fewer nanomaterials are required to produce the same result and the materials last longer. Some estimates suggest the technology can save automobile manufacturers up to $400 per vehicle by lowering the amount of platinum used in catalytic converters.

WHAT TO WATCH FOR   CSI faces stiff competition from Nanophase, Oxonica, and Nanostellar. Also, if fuel cell vehicles gain marketplace acceptance faster than expected, the market for catalytic coatings could reduce significantly.

CONCLUSION   CSI is already selling it products to Honda, GM, and Ford and has established relations with a number of other companies including GE, BASF, and Sumitomo Corporation. Also, as world governments mandate more stringent emission standards, the need for CSI's technology will increase. The technology also has potential to reduce nitrogen oxide that would be of great interest to the power generation industry that creates vast amounts of nitrogen oxide by burning coal.

CSI has successfully completed five different stages of funding and is now well-established in a large market. It appears likely that the company will go public soon. If it does, investors looking for a lower risk investment should consider it.

COMPANY	Eikos, Inc.
INVESTORS	JFC Technologies, Itochu Corp.
ADDRESS	2 Master Drive Franklin, MA 02038
PHONE	508-528-0300
CEO	Joseph W. Piche
WEB	*www.eikos.com*

DESCRIPTION   Eikos develops and licenses unique and highly transparent carbon nanotubes (CNT) for applications in a variety of fields including conductive coatings, circuits, flat panel displays, and solar cells.

WHY IT IS DISRUPTIVE   Eikos conductive ink may be helpful in constructing flat panel displays that are thinner, use less energy, and are more durable. The company has also received a number of grants from the U.S. government—including NASA and the U.S. Air Force—to conduct research and development in the areas of electro-magnetic shielding and transparent coatings. One potential offshoot of such research could be "smart windows"—windows that change their characteristics depending on the need of the user. It could also lead to commercial applications for next-generation cell phones, RFIDs, and flexible electronics.

WHAT TO WATCH FOR   In 2004, Eikos acquired of 40 percent of MysticMD (which develops and licenses conductive medical devices); a move into the medical device arena could open up a lucrative new line of business that might separate it from some of its competitors.

CONCLUSION   Eikos has a strong intellectual property portfolio and has signed promising partnership and licensing arrangements. The market for flat panel displays remains the big prize for the producers of CNTs. If Eikos can license its technology to a big manufacturer, it will be a promising sign. Competition in the field of carbon nanotubes is fierce. Eikos will need to demonstrate it can produce CNTs at scalable levels and affordable prices that offer customers enough of an advantage in terms of cost, ease of use, and product improvement to convince the manufacturers of flat panel displays to switch.

COMPANY	Evident Technologies
INVESTORS	Private
ADDRESS	216 River Street, Suite 200 Troy, NY 12180
PHONE	518-273-6266
CEO	Dr. Clinton Ballinger
WEB	*www.evidenttech.com*

DESCRIPTION   Evident Technologies is a manufacturing and application company that specializes in the production of semiconducting nanocrystals (also known as quantum dots) that have potential applications in the life sciences industry in cell biology, drug discovery, and cancer research.

WHY IT IS DISRUPTIVE   Evident's quantum dots are being produced, marketed, and sold to researchers for biotechnology research. The quantum dots are better than the traditional organic flurophores dyes that are also commonly used to locate, quantify, and identify proteins and genes. The company's nanocrystals are more stable, come in a variety of colors, and last longer. These attributes allow researchers to perform more tests, see cells and other biological samples in greater detail, and provide researchers with the freedom to perform longer-term imaging. Quantum dots could also be used to brighten LEDs and thus replace conventional lightbulbs. The energy cost savings and the concomitant environmental benefits of replacing incandescent and other low-efficiency lighting sources could also reduce the amount of electricity used for lighting by 35 percent (which would have huge implications for the electricity generation industry). Semiconducting nanocrystals might also have applications in telecommunications (optical switches), computing (optical transistors), and photovoltaics (more efficient solar cells).

WHAT TO WATCH FOR   In early 2005, it received a grant from the National Science Foundation to develop advanced anticounterfeiting materials based on its quantum dot technology. If the technology is successful, it could create a very nice niche.

continued

Evident Technologies continued

CONCLUSION   Evident is already selling its product and appears to be positioning itself for growth. It has a strong IP position and a good relationship with Rensselaer Polytechnic Institute, a world leader in nanomaterials. The company faces strong competition from Quantum Dot Corporation and other competitors. The fact that it has not established any relationships with the big biotechnology leaders (such as Quantum Dot Corp has done with Genentech) is a cause for some concern but look for it to rectify this by establishing a relationship with a large biotech leader in the biodiagnostic arena. In 2005, the company also established a partnership with Konarka (see page 214) to explore the use of its quantum dots in flexible polymer photovoltaics. The partnership could lead to the creation of plastic packaging materials that can change their properties to maintain product content longer and keep it fresher.

COMPANY	Five Star Technologies
INVESTORS	Morgenthaler Ventures, Chevron Technology Ventures, Industrial Technology Ventures, Early Stage Partners
ADDRESS	21200 Aerospace Parkway Cleveland, OH 44142
PHONE	440-239-7005
CEO	James Mazzella
WEB	*www.fivestartech.com*

DESCRIPTION   Five Star Technologies has developed and patented a high-through-put technology—called Controlled Flow Cavitation (CFC)—that is capable of producing unique and uniform nanomaterials in large quantities.

WHY IT IS DISRUPTIVE   Five Star's CFC technology is sufficiently different from its competitors to give it an advantage in important niche areas, including wastewater treatment, emission control catalysts, fuel cells, nutraceuticals, and beauty products.

WHAT TO WATCH FOR   Five Star faces stiff competition from Oxonica, Nanophase, Inframat, and Nanotechnologies, Inc. If the market for its materials is limited to the low-margin end uses (like plastics), its prospects are less bright.

CONCLUSION   The fact that Five Star must operate under confidentially agreements makes it hard to evaluate the extent to which its products are being incorporated into the commercial marketplace. However, because its CFC process is scalable and capable of creating uniform nanoparticles (a highly desirable characteristic, especially in the pharmaceutical and drug delivery areas), its prospects look good. This is especially true if its nanomaterials are utilized for drug delivery.

COMPANY	Hybrid Plastics
INVESTORS	Private
ADDRESS	18237 Mount Baldy Circle Fountain Valley, CA 92708
PHONE	714-962-0303
CEO	Dr. Joseph Lichtenhan
WEB	*www.hybridplastics.com*

DESCRIPTION   Hybrid Plastics is the world's leading developer and manufacturer of a unique series of nanostructured hybrid inorganic-organic chemicals called Polyhedral Oligomeric Silsesquioxanes (POSS).

WHY IT IS DISRUPTIVE   POSS can be incorporated into any number of plastics to give it new and improved properties such as lightness, hardness, and heat resistance. The company already has a number of existing customers including Intel, Boeing, Bausch & Lomb, and L'Oreal. Furthermore, it has a strong working relationship with the U.S. government and received an $850,000 grant from the Missile Defense Agency to develop coatings capable of protecting electronic and optical components from radiation. The latter opens a host of space-related applications for its materials.

WHAT TO WATCH FOR   Hybrid Plastics is working with the U.S. Navy to produce materials that would be capable of replacing a variety of metal structures on the topside of its ships. If the company's technology is certified by the Defense Department, it will be another bullish sign for the company.

CONCLUSION   Hybrid Plastics has demonstrated the capability to scale its production to meet the demands of its increasing customer base, and because POSS molecules require no special equipment or processes to be added to existing manufacturing processes, the potential market for its nanomaterials is massive. Given the company's strong IP portfolio and excellent management and scientific team, as well as its strong relations with the U.S. government, if it were to go public, it would make a good addition to any nanotechnology portfolio. Hybrid Plastics will likely face increased competitive pressure from other nanomaterials producers such as Nanocor, Eikos, Nanophase, 3M, and BASF.

COMPANY	Inframat Corporation
INVESTORS	Not available at press time
ADDRESS	74 Batterson Park Road Farmington, CT 06032
PHONE	860-678-7561
CEO	David Reisner, PhD
WEB	*www.inframat.com*

DESCRIPTION Inframat specializes in developing various nanoceramics and nano-materials with unique properties.

WHY IT IS DISRUPTIVE Inframat's nanomaterials can lead to significant improvements in a variety of industrial uses. The U.S. Navy has used Inframat's nanocoatings to reduce the effects of corrosion, keep barnacles from adhering to the hulls of ships, and make other materials that are more flexible, heat-resistant, and/or durable than conventional materials. Its superfine tungsten carbide hard metals—which are sold under the trade names Nanalloy and Infralloy—may make ideal materials for cutting and drilling equipment, and its patented nanocatalysts may help significantly detoxify a variety of volatile compounds—including arsenic.

WHAT TO WATCH FOR If Inframat's plasma spray technology is licensed by additional industrial users (e.g., GE, Rolls-Royce, or Pratt Whitney), it could lead to significant royalty revenues. If, however, the technology is not successful, the nanomaterials side of its business may generate only modest returns—particularly if other nanomaterials manufacturers develop competing materials and succeed in driving the price of those nanomaterials to commodity price levels.

CONCLUSION Inframat's revenues have been growing consistently since 2000, and its relationship with the U.S. military makes it likely that government funding will continue. It has also established relations with industry leaders such as GE and Raytheon and because its technology requires no serious retrofitting on behalf of the end user, it is likely to find more customers. In addition, Inframat has also developed a patent process for distributing nanoparticles uniformly across industrial surfaces. This plasma spray technology makes it possible to coat various ship and aircraft engine parts to reduce wear and tear and allow them to run hotter and more efficiently.

COMPANY	Nanofilm
INVESTORS	Carl Weiss
ADDRESS	10111 Sweet Valley Drive Valley View, OH 44125
PHONE	800-883-6266
CEO	Scott Rickert
WEB	*www.nanofilm.cc.com*

DESCRIPTION   Nanofilm develops and commercializes nanocoatings for a variety of industries including optics, sports, displays, photonics, and military applications. Founded in 1985, it is one of the older nanotechnology companies in the United States.

WHY IT IS DISRUPTIVE   Nanofilm's nanocoatings impart functional benefits to the surfaces of a lot of different products. Its products are already being used to manufacture sunglasses and ski goggles that don't fog, glass lenses that don't scratch, and windshields capable of repelling rain, ice, and snow.

WHAT TO WATCH FOR   Competition from Nanophase, Oxonica, and other nanomaterials suppliers.

CONCLUSION   Nanofilm has experienced double-digit growth every year since 2002. Given the company's growing international presence, there is no reason why it should not be able to continue that trend. There have been rumors that the company might go public in 2005 or 2006. If it does, investors looking for a solid, long-term investment should seriously consider it.

COMPANY	NanoGram Corporation
INVESTORS	NEC, Dow, Asea Brown Boveri (ABB), Venrock Associates, Nth Power Technologies, Harris & Harris, Bay Partners, Rockport Capital, SBV Venture Partners
ADDRESS	2911 Zanker Road San Jose, CA 95134-2125
PHONE	408-321-5001
CEO	Timothy Jenks
WEB	*www.nanogram.com*

DESCRIPTION   NanoGram is an intellectual property holding company founded to develop and commercialize new nanoscale materials for medical, optical, electronic, and energy storage applications and products. It has created three separate entities: NeoPhotonics (which is covered in greater detail on page 233), Kainos Energy Corporation, and NanoGram Devices Corporation—which was acquired by Wilson Greatbatch for $45 million in March 2004.

WHY IT IS DISRUPTIVE   Kainos Energy is working to develop a process for lowering the cost and improving the performance of solid oxide fuel cells. If it is successful, the price and effectiveness of fuel cell technology could make it a more viable energy alternative.

WHAT TO WATCH FOR   Kainos faces a lot of competition in the fuel cell market from other start-ups like NanoDynamics (see page 220), but if either NeoPhotonics or Kainos announce strategic partnerships or licensing agreements with major producers in the telecommunications or energy industries, it will be a bullish sign.

CONCLUSION   NanoGram has strategic partnerships with NEC, Dow, and ABB, and it has received some government funding to help develop its solid oxide fuel cells. NanoGram also has a strong management team and solid intellectual property portfolio. The acquisition of NanoGram Devices by Wilson Greatbatch serves to validate its business model. Investors interested in either NeoPhotonics or Kainos are encouraged to look at investing in Harris & Harris (NASDAQ: TINY), which has a small equity stake in NanoGram.

COMPANY	Nanolab, Inc.
INVESTORS	Angel investors
ADDRESS	55 Chapel Street Newton, MA 02458
PHONE	617-581-6747
CEO	David Carnahan
WEB	*www.nano-lab.com*

DESCRIPTION   Nanolab is one of the earliest producers of nanomaterials, specializing in the production of carbon nanotubes, nanoparticles, and nanowires.

WHY IT IS DISRUPTIVE   Nanolab claims it can produce carbon nanotubes of predetermined length, diameter, and quality. It also purports to have reduced the price of these high-quality carbon nanotubes from $1,000 per gram to around $100. If it can produce high-quality CNTs with the specified properties that customers are demanding at an affordable price, it should be able to find a host of uses for its materials. In fact, one potential application that the U.S. Army is exploring for Nanolab's carbon nanotubes is to help bolster the strength of body armor for military vehicles and tanks.

WHAT TO WATCH FOR   If Samsung or Boeing increase their orders, it will be a bullish signal.

CONCLUSION   Nanolab was founded in 2000, and although quite small, it is one of the few nanomaterials companies to have made a profit. Some of it clients include Boeing, Samsung, and Raytheon, and it has an excellent relationship with the U.S. Army and has received a number of Small Business Innovation and Research (SBIR) grants. Only 25 percent of the company's revenues, however, come from commercial sales—75 percent is from research and development grants. Also, it is only producing pilot-scale amounts of nanomaterials. If it is to survive, it will need more commercial users and/or production partners. Nanolab will face stiff competition from both large industrial producers of carbon nanotubes, such as Frontier Carbon, and smaller competitors such as Carbon Nanotechnologies (see page 83) and Eikos (see page 86).

COMPANY	Nanoledge
INVESTORS	CDP Capital Technology Ventures, Emertec, SORIDEC
ADDRESS	Cap Alpha Avenue de l'Europe Montpellier, Clapiers 34940 France
PHONE	33.4.67.59.36.58
CEO	Pascal Pierron
WEB	*www.nanoledge.com*

DESCRIPTION   Nanoledge designs and develops multifunctional materials out of carbon nanotubes.

WHY IT IS DISRUPTIVE   Nanoledge has an established relationship with Babolat, the French tennis racket manufacturer, to use the company's carbon nanotubes in the production of one of its tennis rackets. The racket is stronger and lighter than today's top-of-the-line rackets and is being used by some of the top professional tennis players. If the company's carbon nanotubes can be incorporated into other products that can take advantage of its properties, a host of material suppliers in the graphite, aluminum, and steel businesses could find the demand for their material decrease.

WHAT TO WATCH FOR   Nanoledge is currently selling its carbon nanotubes at a loss to Babolat—in an attempt to increase market visibility. It also has no additional customers at this time. The company will need to prove it can produce high-quality carbon nanotubes in bulk quantities at affordable prices if it wants to compete internationally with the likes of Carbon Nanotechnologies, SouthWest Nanotechnologies, Zyvex, InMat, and Eikos.

CONCLUSION   Given this state of competition, it is difficult to imagine how Nanoledge can grow much beyond a very small niche player.

COMPANY	NanoProducts Corporation
INVESTORS	Cabot Microelectronics, Southern California Gas Company, Hosokawa Micron Corporation
ADDRESS:	14330 Longs Peak Court Longmont, CO 80504
PHONE	970-535-0629
CEO	Tapesh Yadev
WEB	*www.nanoproducts.com*

DESCRIPTION   NanoProducts designs, develops and manufactures nanomaterials.

WHY IT IS DISRUPTIVE   Unlike so many other nanomaterials companies, NanoProducts does not simply focus on large-scale volume production, it tailors its nanoparticles to meet specific applications. The company has reportedly manufactured over 200 different designs, including everything from nanomaterials for self-cleaning surfaces and polishing slurries for the semiconductor industry to its main product line, PureNano, which is being used by PPG Industries to make automotive coatings that are scratch resistant.

WHAT TO WATCH FOR   NanoProducts will face competition from mid-size competitors like Nanophase and Oxonica. The effect of this competition could be that NanoProducts' nanomaterials face severe downward price pressure. If, however, the company can continue to develop nanomaterials for niche applications—like nanomaterials for biomedical sensors—it could lead to significant profits.

CONCLUSION   The company has been in business for over 10 years and already has a large and diverse customer base as well as established relationships with PPG Industries, Cabot Microelectronics and Sigma Aldrich. Furthermore, it has a very strong IP portfolio. If NanoProducts goes public, it is one of the few nanomaterials companies' investors should consider adding to their portfolio.

Privately Held Nanomaterial Companies

COMPANY	Nanotechnologies, Inc.
INVESTORS	Harris & Harris, Air Products and Chemical (APD), Techxas Ventures, Castletop Capital, Convergent Investors VI, Capital Conceptions
ADDRESS	1908 Kramer Lane
	Austin, TX 78758
PHONE	512-491-9500
CEO	Randy Bell
WEB	*www.nanoscale.com*

DESCRIPTION   Nanotechnologies, Inc. has developed a proprietary pulsed plasma technology capable of producing high-performance nanoparticles.

WHY IT IS DISRUPTIVE   The company's nanoparticles have a variety of applications including transparent coatings (for eyewear), electronic materials, antimicrobial products, fuel additives, and energetic materials (explosives and propellants).

WHAT TO WATCH FOR   If the company can continue to announce new partnerships with established companies, it would serve as a validation of its technology. If it can increase its production quantities, it might allow the company to go public.

CONCLUSION   Nanotechnologies, Inc. has done all the right things. It has partnered with Essilor, a leading manufacturer of eyeglass lenses, and received an equity investment and signed a joint development agreement with Air Products and Chemical (NYSE: APD). It also has an experienced management and scientific team and owns valuable intellectual property. Nanotechnologies, Inc. could make an attractive acquisition target for a company like Air Products and Chemical, but it faces strong competition from companies like Five Star Technologies, Nanophase, and Nanofilm.

COMPANY	Qinetiq Nanomaterials Ltd.
INVESTORS	Wholly owned subsidiary of Qinetiq, of which the U.S.–based Carlyle Group has a 34 percent stake
ADDRESS	Ively Road, Cody Technology Park Hampshire GU14 0LX United Kingdom
PHONE	44(0).1252.393000
CEO	Sir John Chisholm
WEB	*www.nano.qinetiq.com*

DESCRIPTION Qinetiq Nanomaterials is Europe's leading manufacturer and supplier of nanopowders and serves a variety of clients in fields as diverse as explosives and portable power sources.

WHY IT IS DISRUPTIVE Qinetiq's nanomaterials have potential applications in batteries, fuel cells, ceramics, composites, electronics, energetics, sensors, magnetics, and the material sciences. In 2005, the company announced that, in partnership with Intel Corp., it had created a new material that yielded a threefold increase in transistor performance without consuming any additional power. Equally important, the company is capable of getting its products to the market quickly and in bulk quantities.

WHAT TO WATCH FOR The company has yet to crack the U.S. market, although Qinetiq's recent acquisitions of U.S.–based Foster-Miller and Westar Aerospace & Defense Group, Inc. may change that. Like all other nanomaterials companies, it faces tough competition.

CONCLUSION As a subsidiary of Qinetiq, the company can draw on the knowledge and expertise of its 9,000 person staff—including the company's deep pool of scientists and other technical experts. It can also piggyback upon Qinetiq's existing relationships in the marine, energy, telecommunications, automotive, electronics, and defense industries. As a former spin-off from Great Britain's Defense Evaluation Research Agency (the equivalent of the U.S. DARPA), it will also have access to the United Kingdom's defense market. There has been talk of a future IPO for Qinetiq. While Qinetiq Nanomaterials would comprise only a portion of the company's value, it may be worth considering for investors looking to diversify their portfolio.

Privately Held Nanomaterial Companies

COMPANY	Quantum Dot Corporation
INVESTORS	Versant Ventures, Abingworth Management, Technogen Associates, Schroder Ventures, Frazier & Co., MPM Asset Management, CMEA Ventures
ADDRESS	26118 Research Road
	Hayward, CA 94545
PHONE	510-887-8775
CEO	George W. Dunbat
WEB	*www.qdots.com*

DESCRIPTION   Quantum Dot Corporation is a world leader in the development, marketing, and sale of novel semiconductor nanocrystals (quantum dots).

WHY IT IS DISRUPTIVE   Quantum dots, due to their small size and unique coloring properties, can be used to bind to other molecules that then allow researchers to better track targets of interest. Quantum dots also show great potential in cancer tumor staging by allowing doctors to determine where and to what degree cancer cells have spread. The technology has the potential to revolutionize both the medical diagnostic and the medical imaging industries.

WHAT TO WATCH FOR   Competition in the diagnostic arena is fierce, and a number of other companies are working to develop faster, cheaper, and more sensitive and accurate diagnostic tools. Investors need to be aware of two additional concerns. One, questions still remain about the toxicity and environmental impact of semiconducting nanoparticles and nanocrystals; if they are not given a clean bill of health, their value could plummet. Second, even if quantum dots are successful, the pressure to contain health care costs may make the government and insurance companies reticent about adopting the technology if it is not priced accordingly.

CONCLUSION   The company has already provided its products to thousands of researchers, and it is collaborating with a number of pharmaceutical companies. The company also has a very strong intellectual property portfolio and controls some of the key patents for biotech applications. If Quantum Dot can achieve profitability in 2005 and continue to line up additional corporate partnerships, investors are encouraged to consider an investment if the company goes public.

COMPANY	QuantumSphere, Inc.
INVESTORS	Private Angel investors
ADDRESS	1041 West 18th Street, Suite B102 Costa Mesa, CA 92627
PHONE	949-574-3000
CEO	Kevin Maloney
WEB	*www.qsinano.com*

DESCRIPTION   QuantumSphere is a manufacturer of metallic nanopowders for the cosmetic, defense, aerospace, and automotive sectors.

WHY IT IS DISRUPTIVE   Although quite similar in nature to Catalytic Solutions and Nanotechnologies, Inc., QuantumSphere claims that it has completed a large-scale reactor for the production of nanonickel. Due to nanonickel's unique properties, it is possible that the material could replace platinum in catalytic converters as well as in fuel cells. In 2005, the company announced it was supplying nanonickel to Dermacia for its advanced line of therapeutic cosmetics for severe acne sufferers.

WHAT TO WATCH FOR   If QuantumSphere can achieve large-scale production, its nanonickel has the potential to find a lucrative market, but the company faces stiff competition. If the company is to thrive, it needs to find alternative uses for some of its other nanomaterials. For instance, nanoaluminum could be used to achieve faster burn rates in rockets, and still other nanomaterials might make for higher yielding explosives. Like all other nanomaterials, QuantumSphere's materials could experience rapid price depreciation as other companies scale up their production of competitive materials.

CONCLUSION   The company has no announced plan to go public, but if it does, investors are encouraged to seek additional confirmation of nanonickels alleged superiority before investing.

Privately Held Nanomaterial Companies

COMPANY	SouthWest Nanotechnologies, Inc.
INVESTORS	Private
ADDRESS	2360 Industrial Boulevard Norman, OK 73069-8518
PHONE	405-217-8388
CEO	Howard G. Barnett
WEB	*www.swnano.com*

DESCRIPTION   SouthWest Nanotechnologies specializes in the development and manufacture of single wall carbon nanotubes (SWNTs).

WHY IT IS DISRUPTIVE   SouthWest has received a $600,000 SBIR grant from NASA to produce extremely strong, durable, and light materials for space-based applications. Specifically, the nanotubes could lead to a reduction in the weight of spacecraft—which, in turn, could lead to cost savings. (It is estimated that it costs NASA approximately $10,000 to launch one pound of material into space.) If successful, SouthWest's material could be incorporated into aerospace and automotive applications.

WHAT TO WATCH FOR   The key to success for the company appears to be incorporation of its SWNT into the aerospace market. If it can enter this market through a partnership or a licensing arrangement with a company like Boeing, it would be a very bullish sign. Also, if its SWNT are licensed to a manufacturer of flat panel displays, it would represent another large, lucrative market.

CONCLUSION   SouthWest Nanotechnologies appears to have developed a niche in the field of SWNTs. It remains to be seen whether this specific market is sufficiently distinct from what the other producers of carbon nanotubes are developing. The fact that the company has established a relationship with ConocoPhillips and is one of the few carbon nanotube manufacturers certified by Zyvex—a leading supplier of nanotechnology products and tools (see page 241)—to have its materials used in its carbon nanotube additive NanoSolve is a positive sign. The company also has a strong intellectual property portfolio, especially in the area of SWNT production (CoMoCat process). Like every other carbon nanotube manufacturer, it will face fierce competition, and it may need to partner with a larger company to produce the quantities necessary to survive in the commercial marketplace.

COMPANY	Starfire Systems.
INVESTORS	Harris & Harris and others
ADDRESS	10 Hermes Road, Suite 100 Malta, NY 12020
PHONE	518-899-9336
CEO	Richard M. Saburro
WEB	*www.startfiresystems.com*

DESCRIPTION   Starfire System develops and manufactures high performance silicon carbide ceramic products and materials for electronics, aerospace, and material science applications.

WHY IT IS DISRUPTIVE   Starfires proprietary liquid polymers create nanostructured ceramics with superior properties. For instance, the company's composite brakes are not only less likely to wear out and quieter, they are also substantially lighter than traditional brakes and reduce the amount of energy needed to accelerate the automobile—thus reducing fuel consumption and increasing the acceleration rate. Some of its other materials function better in high temperature, corrosive, and high-wear environments.

WHAT TO WATCH FOR   NASA has funded Starfire to supply materials to help build vehicles that might be used on the moon or, possibly, Mars. If successful, it'll serve as further validation for its technology.

CONCLUSION   Starfire will need to continue to break through cost, performance, and price barriers in order to grow. Investors interested in the company are encouraged to consider an investment in Harris & Harris (see pages 175–76).

	COMPANY	Triton Systems, Inc.
	INVESTORS	Millennium Materials Technology Fund (Only Triton BioSystems)
	ADDRESS	200 Turnpike Road Chelmsford, MA 01824
	PHONE	978-250-4200
	CEO	Samuel Straface
	WEB	*www.tritonsys.com*

DESCRIPTION   Triton is an applied research and development company focused on creating products for government and commercial markets. It has spun-off four separate nanomaterials businesses: Triton BioSystems, Inc., Sensera, Inc., SI2 Technologies, and Elecon.

WHY IT IS DISRUPTIVE   Triton BioSystems's Targeted Nano-Therapeutics (a nanoparticle cancer treatment) combines a magnetic nanoparticle with an antibody. The antibody then selectively attaches itself to a cancer cell, and the magnetic nanoparticle is heated to a temperature high enough to kill the cancer cell. If effective, it would represent a significant improvement over chemotherapy and/or radiation treatment.

WHAT TO WATCH FOR   Triton Biosystems Targeted Nano-Therapeutics nanoparticle cancer treatment technology is the company's crown jewel. If it receives FDA approval, it would be a significant milestone. The other companies under Triton's umbrella also bear watching, but each faces real competition. For instance, Sensera is up against companies like Quantum Dot in the point of care diagnostic market, and SI2 Technologies and Elecon have targeted the highly competitive electronics and semiconductors markets.

CONCLUSION   The structure of Triton Systems makes it difficult to follow. However, the parent company has licensed some of its nanomaterials to Ashland Chemical Company (although the terms and size of the deal are unknown), and another of its nanomaterials has been approved by the FDA as a MRI imaging agent. This suggests that the company's management has the knowledge and expertise to successfully navigate new products to market. Any cancer treatment, however, can expect a long, rigorous and costly regulatory review before it reaches the commercial marketplace.

COMPANY	Hyperion Catalysis International, Inc.
INVESTORS	Private
ADDRESS	38 Smith Place Cambridge, MA 02138
PHONE	617-354-9678
CEO	Samuel Wohlstadter
WEB	*www.hyperioncatalysis.com* or *www.fibrils.com*

DESCRIPTION   Hyperion Catalysis is arguably one of the oldest nanotechnology companies in existence, having been around since 1982. The company develops, manufactures, and supplies carbon nanotubes—which it calls FIBRILs.

WHY IT IS DISRUPTIVE   The multiwalled carbon nanotubes (MWNTs) have been supplied for some time to resin companies to make plastic components with enhanced capabilities for the automotive industry. Specifically, its products allow for the creation of electroconductive polymers, which in turn allow automotive companies to use existing electrostatic paint systems (one of the more expensive aspects of building a new car) on the exterior of the body.

WHAT TO WATCH FOR   Hyperion is attempting to produce flame retardant plastics as well as CNT-enhanced plastic antennas that can be used for high-speed wireless communication devices. Both markets are higher margin than its existing markets and may bring greater profits.

CONCLUSION   Hyperion has a long track record, and it has existing relationships with industry leaders such as GE Plastic and Degussa. The fact that the market for its products is expected to grow 20 percent annually through 2008 bodes well for its future prospects. The company also has a strong intellectual portfolio with over 100 patents. The biggest problem it faces is that the market for carbon nanotubes for plastic components is likely to become a low-margin business. It also faces a number of competitors like Nanocor, Southern Clay, Zyvex, and Carbon Nanotechnologies.

COMPANY	Nanomix, Inc.
INVESTORS	Seven Rosen Funds, Apax Partners, North Star Equity, EnerTech Capital Partners, Harris & Harris
ADDRESS	5980 Horton Street Emeryville, CA 94608
PHONE	510-428-5300
CEO	David Macdonald
WEB	*www.nano.com*

DESCRIPTION   Nanomix is a leader in the development of carbon nanotube-based nanoelectronics sensors.

WHY IT IS DISRUPTIVE   Nanomix's sensors have a wealth of industrial applications in the chemical and gas sensor market. For instance, the gas and oil industry could employ such sensors throughout a refinery for industrial process control and/or to monitor gas leaks and air quality. Other customers could use the sensors for pollution or pathogen detection, chemical detection in drinking water, and indoor air quality sensing in schools. Nanomix has already successfully produced a medical capnography sensor (a small, inexpensive, easy-to-use device that can quickly and accurately detect if carbon dioxide is present). The company has also successfully licensed field-emitting thin film materials containing carbon nanotubes to Dupont Electronic Technologies, which could lead to high-quality, affordable flat panel televisions.

WHAT TO WATCH FOR   In April of 2005, Nanomix announced the creation a hydrogen sensor system. If successful, it could represent a very significant and profitable long-term market—especially if fuel cell technology becomes prevalent. Longer term, investors should watch to see if the company is able to develop a sensor that mimics the human sensory system, including an artificial nose and tongue.

CONCLUSION   Nanomix's sensors appear to have superior price/performance compared to traditional technologies, and its intellectual property portfolio, strong management, and scientific teams all suggest that it well positioned for future growth. The company has not yet received any significant revenue from its agreement with Dupont nor has its medical capnography sensor been widely commercialized.

Privately Held Nanomaterial Companies

COMPANY	Nanoscale Materials, Inc.
INVESTORS	Private
ADDRESS	1310 Research Park Drive Manhattan, KS 66502
PHONE	785-537-0179
CEO	Bill Sanford
WEB	*www.nanmatinc.com*

DESCRIPTION   Nanoscale Materials is a developer and producer of nanomaterials that have been especially designed to attack and neutralize chemical, biological, and other hazardous materials and agents.

WHY IT IS DISRUPTIVE   In 2003, the company's nanomaterials were sold to make masks that were created to address the SARS virus. Its lead product, FAST-ACT (First Applied Sorbent Treatment Against Chemical Threats) decomposes toxic chemicals and could be an important tool in the event of a chemical or biological attack.

WHAT TO WATCH FOR   If the company can develop nanomaterials that neutralize hazardous waste or if it licenses its technology to a major cloth producer for the purpose of creating germ or bacteria-neutralizing clothing, the company could gain access to some large and potentially lucrative markets.

CONCLUSION   Nanoscale Materials has demonstrated successful sales of its Nano-Active materials and FAST-ACT. And although the need for its product in addressing chemical and biological agents are small at the present time, they are invaluable and would be in demand in a crisis.

Privately Held Nanomaterial Companies

COMPANY	Nanostellar
INVESTORS	3i Group, William Miller, Frank Marshall
ADDRESS	3603 Haven Avenue, Suite A Menlo Park, CA 94025
PHONE	650-368-1010
CEO	Michael Pak
WEB	*www.nanostellar.com*

DESCRIPTION   Nanosteller uses a proprietary software program to develop highly efficient nanomaterial catalytic solutions that reduce the amount of platinum necessary for controlling automobile emissions and for generating energy from fuel cells.

WHY IT IS DISRUPTIVE   Environmental and health concerns are likely to continue to drive the demand for tougher emission standards. Platinum, which is quite expensive, is the most commonly used material for catalytic converters, but it can add up to $400 to the price of an average automobile. Nanostellar's technology promises to be able to reduce that sum considerably.

WHAT TO WATCH FOR   If Nanostellar can successfully license its software to another nanomaterials producer, it may represent an alternative—and ultimately more sustainable—business model than producing the materials themselves. The company might also make a strategic acquisition target for a company like Symyx, who has already developed a specialty in manufacturing software to create new nanomaterials solutions.

CONCLUSION   Nanostellar's software programs appear to be unique and could offer a competitive advantage over its competitors. Moreover, the company's management team and investors are scientifically knowledgeable, well networked, and have a great deal of experience in the commercial marketplace. The problem is that the marketplace for catalytic solutions is tough. Catalytic Solutions and Oxonica already have products on the market, and bigger players like Dow and ExxonMobil are also actively engaged in the market. Unless Nanostellar's software develops nanomaterials that are significantly better than its competitors, the company's late start and limited resources will make it difficult for it to compete.

Summary

Like the equipment and tools discussed in Chapter 3, there is little doubt nanomaterials will be an integral component of the success the field of nanotechnology achieves. But the profits are likely to be even more modest than the nanotech equipment and tool sector. As a general rule, I am bearish on the entire nanomaterials markets for the following four reasons:

1. As nanomaterials manufacturing expertise continues to develop, prices will be driven down to commodity levels, and profit margins will be squeezed ever tighter. As a result, the advantage will go to bigger companies, such as Dow Chemical, BASF, and Degussa (which are discussed in Chapter 5), who have the ability to produce these nanomaterials in bulk.

2. Most of the value of nanomaterials resides not in the material per se but rather in how that material can increase the performance of the end product. Therefore, investors are encouraged to focus less on the materials themselves and more on how they are being applied to create competitive advantages for the end users.

3. Intellectual property issues continue to hang like a heavy cloud over much of the field—particularly in the area of carbon nanotubes. Until such issues are resolved, investors risk having some of their money tied up in companies who are spending more of their time in the courtroom instead of pushing their products into the commercial marketplace.

4. Lastly, investors need to monitor environmental issues regarding the health and environmental aspects of nanomaterials. Although no nanomaterials have yet been found harmful, neither have they been given a clean bill of health. Until they receive such a clearance, investors need to be cautious.

"Nanotech is absolutely critical to where we want to go."

—Jeff Immelt, CEO of General Electric

Chapter 5

It's a Dog-Eat-Dog World: The *Fortune* 500 Companies

In his book, *Quantum Investing*, Stephen Waite suggests that by 2025 at least half of the companies comprising today's Dow Jones Index will have been replaced. He bases his prediction on his understanding of quantum physics, which he notes is the underlying science behind 30 percent of today's Gross National Product, and suggests its influence will only grow more pervasive by 2025 and thus play a role in removing some of today's leading companies from the Index.

It is a thoughtful proposition, but it begs the question: Which companies will survive, and which are fated to go the way of Tennessee Coal & Iron and, more recently, International Paper and AT&T?

Waite offers the first clue when he writes that "the mother of all quantum revolutions—the nanotechnology revolution—has the potential to be to the twenty-first century what microelectronics was to the twentieth century." The second clue can be found in the findings of Lux Research—the country's preeminent nanotechnology research firm. In an early report it noted that thirteen of the thirty companies listed on the Dow mentioned nanotechnology on their Web sites (the list is now up to eighteen).

While it is not my intention to suggest that the mere mentioning of nanotechnology on one's Web site holds the key to retaining a position on the Dow Index, it does offer an attractive first screen for discerning which will survive because it suggests those companies are at least sensitive to the role nanotechnology will play in defining the twenty-first century.

The eighteen companies mentioning nanotechnology are 3M, Boeing, Proctor & Gamble, Microsoft, United Technologies, SBC Communications, DuPont, Philip Morris, ExxonMobil, GE, General Motors, Hewlett-Packard, Honeywell, IBM, Intel, Johnson & Johnson, Merck, and Pfizer. A cursory review of these companies' involvement in nanotechnology suggests that most of them are approaching the space in a very pragmatic way. They understand nanotechnology has practical applications that can make their existing products better *today*, while also having the potential to fundamentally transform their business *tomorrow*.

It is this balanced approach to applying nanotechnology to improving the company's next quarterly financial statement, together with commitment to nanotechnology research and development that will allow these companies to successfully navigate and stay afloat in the turbulent and uncharted waters of the future.

For instance, 3M has developed a new nanocomposite for tooth filler that is stronger, more durable, and keeps its polish longer. Such advances are the stuff that makes up sustained annual growth. Longer term, the company is looking at nanocomposites for bone regeneration. If it can master this technique, it will have gained entry to a new, large, and potentially high-margin area—the type that might help it stay on the Dow for another two decades.

General Motors is following a similar strategy. It is employing nanocomposites to make a number of its newer vehicles more resilient and fuel efficient—including the Chevy Impala and Hummer H2. Down the road, it is looking to nanotechnology to improve both hybrid battery technology and fuel cell technology—which might help the company survive when the century-long reign of the internal combustion engine finally ends.

DuPont is another company following a similar midterm/long-term strategy. It recently signed a licensing agreement with Nanomix to manufacture flat panel displays using carbon nanotubes. The displays will be flatter and more energy efficient than today's state-of-the-art displays. It is also, however, investing millions into the U.S. Army's Institute of Soldiering Nanotechnologies. Why? Because DuPont knows the Army is interested in producing clothing that can monitor the health of the wearer, change thermal properties, produce its own energy, and potentially even change colors on demand, and if it can stay abreast of such developments, it will give the company a leg up on identifying the next great material that could revolutionize the textile industry.

Hewlett-Packard, Intel, and IBM are investing millions in nanotechnology, and it is why all three have entered into partnerships with some of today's most promising private nanotechnology start-ups, including ZettaCore, Molecular Imprints, Zyvex, and Nanosys. They understand that nanotechnology can improve the performance of computer chips today, help make next-generation integrated circuits in a few years, and, ultimately, usher in the era of either carbon nanotube-based chips or molecular electronics.

The pharmaceutical industry is no different. Johnson & Johnson, Merck, and Pfizer recognize that in the near future a number of their most lucrative patents will expire, and proprietary nanoparticles could help extend those patents. Longer term, they understand that dendrimers and other nanoscale devices might well be the new drug delivery platforms of the future, and if they want to survive, they need to start working on developing those platforms and technologies today.

The very size and diversity of these large companies makes it unrealistic (and imprudent) to evaluate their stock solely on the basis of their involvement in nanotechnology because at the present time nanotechnology-related products make up only a fraction of these companies revenues. The purpose of this chapter is to highlight the exciting work being done in nanotechnology by these and other large companies because much of it has been—and will continue to be—overlooked by financial analysts who are only evaluating these companies on the basis of standard and conventional metrics. Investors, however, who understand how these companies' endeavors in the field of nanotechnology are positioning themselves for future growth might be able to spot some nice opportunities before other investors catch on.

The profiles in this chapter cover only the most relevant nanotechnology developments of the listed companies and do not make an attempt to review every aspect of these large and diversified companies. You are therefore encouraged to consult additional independent analyses of these companies (e.g., Morningstar) for a more comprehensive overview of the companies' stocks short-term prospects.

You should also be aware that there are other companies involved in nanotechnology that are not covered in the book. For instance, Kraft, a division of Philip Morris, has created the NanoteK Consortium—a group of fifteen different government and academic laboratories—to explore everything from how nanotechnology can improve packaging to how it can lead to the creation of "interactive" foods and beverages that can be tailored to individual user's tastes. L'Oreal has been using nanoparticles for over ten years to make its lotions and cosmetics more effective and has one of the largest portfolios of nanotechnology-related patents in the world. Lockheed Martin is partnering with NASA on nanotechnology and is working on a variety of nanotechnology-related initiatives relating to the creation of lightweight, high-strength nanocomposites; nanosensors that can detect problems before they even arise; and even nanomaterials that may one day be able to self-heal. All are worthy of tracking but in the interest of time and space, only those *Fortune* 500 companies working on the most promising nanotechnology-related developments have been included in this chapter.

MMM	COMPANY	3M Corporation
	SYMBOL	MMM
	TRADING MARKET	NYSE
	ADDRESS	3M Center
		Maplewood, MN 55144
	PHONE	888-364-3577
	CEO	Not available at press time
	WEB	*www.3m.com*

DESCRIPTION   3M is a diversified manufacturer of over 50,000 industrial, commercial, consumer, and health care–related products. Best known for its Post-it and Scotch brands, 3M is applying nanoscale developments to a host of products, including optical components, thin film coatings, adhesives, resins, and nanocomposite tooth fillers.

REASONS TO BE BULLISH

► Given the company's excellent history of innovation and the fact that 3M possesses the third largest number of nanotechnology-related patents (after IBM and L'Oreal), it appears well positioned to employ nanotechnology to build on its wide range of existing products as well as manufacture new ones.

► 3M currently invests 6.5 percent of its revenue back into research and development (about $1.1 billion in 2004), suggesting its future prospects for innovation remain promising.

► It possesses a strong management team, excellent brand recognition, and a worldwide manufacturing presence to take advantage of new developments.

REASONS TO BE BEARISH

► 3M's large size and the diversity of its products make market-beating returns difficult (especially given that it has already achieved impressive growth the past three years).

► It faces some potentially damaging asbestos litigation.

► The company is not currently well positioned in the higher-growth, higher-margin display and graphics segment.

continued

3M Corporation continued

WHAT TO WATCH FOR   3M's pharmaceutical business—especially its drug delivery systems—have great potential. The company is also working on nanoscale composites that mimic human bone and may be applicable to bone regeneration. If it can move into either area, it will be a bullish sign, as would the development of nanoscale devices and/or agents used for the detection and treatment of biochemical agents.

CONCLUSION   Outlook is bullish. Under former CEO James McNerney's leadership, 3M has cut costs and returned to growth. The fact that it once again focused on research and development—particularly on nanoscale developments—suggests that it is capable of continued growth, especially if it can develop nanotechnology-related products for the higher-margin health care areas.

AMD	COMPANY	Advanced Micro Devices
	SYMBOL	AMD
	TRADING MARKET	NASDAQ
	ADDRESS	One AMD Place Sunnyvale, CA 94088
	PHONE	408-749-4000
	CEO	Hector de J. Ruiz, Ph.D.
	WEB	*www.amd.com*

DESCRIPTION   Advanced Micro Devices (AMD) is the world's second largest producer of microprocessors and, quite likely, the largest supplier of flash memory. Revenues are split fairly evenly between the two.

REASONS TO BE BULLISH

► In 2004, revenues increased substantially (42 percent) to $5 billion, and profits were $91 million (although, in the first half of 2005, revenues were off last year's pace by 1 percent).

► In 2003, AMD purchased Coatue, a private nanotechnology start-up company developing nonvolatile plastic memory that reportedly has fifteen times the storage density of current flash memory (32Gbit) and was fifteen times cheaper. The development of this technology could possibly lead to a number of applications, including instant-on computers, integrated PCs on a single chip, solid state video storage, ultrafast hard disk drive replacements, as well as new applications in cell phones, mobile devices, and even flexible, plastic identification tags.

► AMD, together with IBM, has announced a breakthrough process that could increase transistor speed up to 24 percent.

► It is partnering with Albany Nanotech—a world-leading facility in terms of developing next-generation nanoelectronics to continue to produce new materials that can increase transistor speed.

continued

Advanced Micro Devices continued

REASONS TO BE BEARISH

► In 2005, AMD announced it intended to spin-off Spansion, its flash memory business, in an IPO. The business accounted for nearly one-half of the company's revenues in 2004.

► Coatue's technology faces stiff competition from Thin Film Electronics (a Swedish company in which Intel has a stake), Hewlett-Packard, and a host of other private start-ups such as ZettaCore.

► In terms of chip production, Intel appears poised to beat AMD to the marketplace with its 45 nm chips by 2007. If so, Intel will likely add some market share.

► AMD's poor cash position (the company is about $1.8 billion in debt) may prevent it from pursuing other promising nanotech-related start-ups.

WHAT TO WATCH FOR   If AMD can produce a superior flash product (e.g., a 32Gbit flash memory product sometime in 2005) or beat Intel to the market with either a 65 nm or 45 nm chip, investors may wish to reconsider.

CONCLUSION   Outlook is bearish. AMD is experiencing great pressure in both of its key markets and faces a hypercompetitive landscape. Intel currently looks better positioned to apply nanotechnology to the development of next-generation computer chips.

AFFX	COMPANY	Affymetrix
	SYMBOL	AFFX
	TRADING MARKET	NASDAQ
	ADDRESS	3380 Central Expressway Santa Clara, CA 95051
	PHONE	888-362-2447
	CEO	Stephen Fodor, Ph.D.
	WEB	*www.affymetrix.com*

DESCRIPTION   Affymetrix (AFFX) is a leader in the field of DNA chip technology and has designed a variety of platforms for analyzing complex information. The company's GeneChip technology makes it possible to simultaneously study the activity of thousands of genes and establish previously unknown links in identifying genetic variations associated with disease.

REASONS TO BE BULLISH
▶ In 2004, AFFX's revenues increased 15 percent to $346 million, and profits were up from $14 to $47 million, although revenues were slightly off this pace for the first half of 2005.
▶ Its GeneChip System has scores of customers in the pharmaceutical, agricultural, and biotechnology industries, and the company presently controls over 70 percent of the world market in the DNA chips.
▶ Its strong intellectual property portfolio (over 250 patents) and the fact that its technology may be useful in developing scores of new drugs to treat prostrate cancer, sudden infant death syndrome (SIDS), Alzheimer's disease, neutral tube defects, Parkinson's disease, and autism suggest it is well positioned for future growth.

REASONS TO BE BEARISH
▶ As technology progresses, the future of health care may reside more in proteomic research and less in the human genome. As such, the development of protein chips may be more important than the DNA chip.
▶ It faces competition from private nanotech start-ups such as BioForce Nanoscience and NanoString.

continued

Affymetrix continued

WHAT TO WATCH FOR   Watch for disruptive technologies that can sequence genomic information even faster and cheaper than Affymetrix can.

CONCLUSION   Outlook is bullish. As Affymetrix's tools become even more affordable, it is likely that more companies and institutions will employ the devices. In turn, this will lead to an increased number of important breakthroughs.

>> **Nanotech in Asia**

In addition to Hitachi and NEC, a number of other companies in Asia are also investing in nanotechnology and are worth watching. For instance, Nissan Motor is now installing lightweight nanocomposites on a number of its vehicles, and Toyota is exploring ways to use carbon nanotubes to store hydrogen in fuel cell cars. Fujitsui and Nippon Telegraph and Telephone Corp. are exploring ways to utilize carbon nanotubes in next-generation computer chips, and Samsung is also using them to construct flat panel televisions. Samsung is even coating some of its home appliances with nanoparticles of silver to prevent and destroy the spread of bacteria.

Mitsui (MITSY), Japan's largest trading company, is investing heavily in nanotechnology and has created both the Bio Nano Research Institute and the Carbon Nanotechnology Research Institute to help develop nanotech-related products. Other companies, including Matsushita and Mitsubishi, are scaling up production of quantum dots, carbon nanotubes, and fullerenes. The former claims to be producing 40 tons of carbon nanotubes, while Mitsubishi—which hopes to use fullerenes in flat panel displays—expects to increase its capacity to 1,500 tons per year and has also created a $100 million venture fund to invest in promising nanotechnology-related companies.

	COMPANY	BASF
BF	SYMBOL	BF
	TRADING MARKET	NYSE
	ADDRESS	Carl-Bosch Strasse 38 Ludwigshafen, 67056 Germany
	PHONE	800-269-2377
	CEO	Dr. Jurgen Strube (Chairman of the Supervisory Board)
	WEB	*www.basf.de*

DESCRIPTION   BASF is one of the world's leading chemical companies and has five separate business segments: chemicals, plastics, agricultural products and nutrition, performance products, and oil and gas.

REASONS TO BE BULLISH

► BASF has a long history of incorporating new technologies to improve the existing products of its customers. Nanotechnology, by its very nature, plays perfectly into this philosophy (i.e., "At BASF, we don't make the product . . . we make the product better).

► The company has invested heavily in nanotechnology and is devoting an estimated 10 percent of its research and development budget to the field and employing a significant number of its 9,000 member R & D staff to developing novel nanomaterials.

► BASF has established a $100 million venture capital fund to invest in promising nanotechnology start-ups and has taken an equity stake Oxonica (see page 80) and Catalytic Solutions (see page 85), and it has an established partnership with Nanophase (see page 78) for the development of new cosmetic sun protection products.

REASONS TO BE BEARISH

► Although it is a significant player in the nanotechnology field, BASF is such a large company that the prospects of significant stock price appreciation are unlikely.

► The chemical industry, in general, is cyclical, and BASF is not immune from broad economic downturns.

continued

BASF continued

WHAT TO WATCH FOR   BASF has announced it is working to develop new nanomaterials capable of storing hydrogen for fuel cells. If successful, it may represent a new and very sizeable market. Also, watch for acquisitions of promising nanotechnology companies like Oxonica. With more than $3 billion in cash, it has resources.

CONCLUSION   Outlook is bullish. BASF is a good, solid company, and due to its emphasis on nanotechnology, investors would be well served to consider an investment in the company. It is working on a number of new nanotech-related products—from novel printing inks that will make plastics scratch resistant to hydroxyapatite that can restore damaged tooth—that will help the company achieve steady but not spectacular growth.

CBT	COMPANY	Cabot Corporation
	SYMBOL	CBT
	TRADING MARKET	NYSE
	ADDRESS	Two Seaport Lane Boston, MA 02210-2019
	PHONE	617-345-0100
	CEO	Kennett F. Burnes
	WEB	*www.cabot-corp.com*

DESCRIPTION Cabot Corp. is a global specialty chemical and materials company. It produces carbon black, fumed silica, inkjet colorants, and, more recently, nanoparticles.

REASONS TO BE BULLISH

► In 2004, revenues increased 11 percent to almost $500 million, and profits were up 17 percent to $34 million. The pace of growth slowed slightly in the first part of 2005.

► Cabot purchased Superior Micropowders, a company specializing in the production of nanopowder catalysts for proton exchange membrane (PEM) fuel cells.

► The company's other products, especially its Nanogel materials, are being used by a variety of companies as well as architects and builders to create windows and sky roofs that have superior insulating, sound-deadening, and self-cleaning properties.

► It has a strong intellectual property portfolio, global presence, and excellent relationships with both manufacturers and distributors, including GE, Pilkington, PLC, and Kalwall Corporation.

REASONS TO BE BEARISH It is unclear if Superior Micropowders technology is superior to that of numerous other nanopowder producers, including Qinetiq, QuantumSphere, or Nanostellar. Also, Aspen Aerogels (see page 81) may give Cabot some stiff competition in the area of aerogels. The entire specialty chemical market is extremely sensitive to macroeconomic forces, and even if its new venture in nanopowders is successful, it is unlikely to significantly affect the company's stock price in the short to midterm.

continued

Cabot Corporation continued

WHAT TO WATCH FOR   Cabot (CBT) should not be confused with Cabot Microelectronics Corporation, which is a separate stand-alone nanotechnology company that is traded on the NASDAQ under the symbol CCMP. (The company produces chemical mechanical planarization (CMP) slurries for the semiconductor industry and is in competition with Nanophase and others.)

CONCLUSION   Outlook is bullish. Cabot is unlikely to experience large gains, but it is a solid company. If the market for PEM fuel cells begins to emerge, Cabot will stand to benefit as a major supplier of nanopowders.

Fortune **500 companies**

CVX	COMPANY	ChevronTexaco
	SYMBOL	CVX
	TRADING MARKET	NYSE
	ADDRESS	100 Chevron Way Richmond, CA 94802
	PHONE	713-752-3854
	CEO	David J. O'Reilly
	WEB	*www.chevrontexaco.com*

DESCRIPTION ChevronTexaco is the second largest U.S. oil company and operates across all segments of the oil and gas industry—exploration, production, refining, transportation, and marketing—and in over 180 countries around the world. It has invested in Konarka (see page 214) and spun off a new nanotechnology subsidiary manufacturing diamondoid materials called MolecularDiamond Technologies.

REASONS TO BE BULLISH

▶ MolecularDiamond Technologies makes and sells higher diamondoids—nanoscale diamond fragments that possess unique properties in terms of surface area, rigidity, and durability—that could potentially be useful in refining processes as well as next-generation optical, electronic, and pharmaceutical devices.

▶ Its equity stake in Konarka—one of the most promising solar cell nanotechnology start-ups—suggests the company is serious about positioning itself as a player in the emerging alternative energy market.

REASONS TO BE BEARISH

▶ ChevronTexaco faces strong competition from ExxonMobil, BP, ConocoPhillips, and Royal Dutch Shell. ExxonMobil is better positioned in the area of employing nanoscale catalysts to refine and process oil and gas, and BP appears to be ahead in its understanding of how nanotechnology may enable fuel cell technology.

▶ It is also subject to the inherit risks of the oil and gas market, namely fluctuating oil prices, weather-related issues, spills, and geopolitical risks (ChevronTexaco has facilities in Angola, Kazakhstan, and Nigeria).

continued

ChevronTexaco continued

WHAT TO WATCH FOR   If MolecularDiamond is successful in finding higher-end market applications for its nanomaterials, it may be worth taking another look at the company. Also, if ChevronTexaco moves aggressively into on-site production of hydrogen for the emerging fuel cell market, investors may want to take another look at this stock.

CONCLUSION   Outlook is bearish. Relative to its oil business, ChevronTexaco's nano-technology ventures are a small fraction of its overall business. Until the company moves more aggressively into either the solar cell market (through its investment with Konarka) or the emerging fuel cell market, investors looking for a more solid energy-related nanotech investment should consider Headwaters or Engelhard.

DGXG.DE	COMPANY	Degussa
	SYMBOL	DGXG.DE
	TRADING MARKET	Frankfurt
	ADDRESS	P.O. Box 30 20 43 40402 Düsseldorf Germany
	CEO	Dr. Utz Hellmuth Felcht
	WEB	*www.degussa.com*

DESCRIPTION Degussa is Germany's third largest chemical company and specializes in producing specialty chemicals for the plastics and automotive industries.

REASONS TO BE BULLISH

► Degussa has clearly identified nanotechnology as an area for future emphasis. In 2000, it created the Degussa Advanced Nanomaterials Center and invested more than $100 million to produce new nanomaterials for a variety of fields, including coatings, catalysts, paints, automobile tires, and filters.

► In 2004, it announced it would be investing 50 million euros to create the Nanotronics Science to Business Center, which is expected to focus on the development of nanomaterials for the electronics industry, lithium ion batteries for the automotive sector, and nanomaterials for the solar cell and radio frequency identification tag markets.

REASON TO BE BEARISH

► In a number of its markets, Degussa will have a hard time distinguishing itself from its competitor's products.

WHAT TO WATCH FOR A number of companies are hoping to reach the flat panel market first with innovative new nanomaterials; look for Degussa to be a strong contender. If it does, it will be another bullish sign.

CONCLUSION Outlook is bullish. Degussa is far more aggressive in its willingness to embrace the potential of nanomaterials. Given its size, strong emphasis on research and development, and excellent rapport with its customers, Degussa is likely to be a leader in the development of new nanomaterials that have real-world applications.

DOW	COMPANY	Dow Chemical
	SYMBOL	DOW
	TRADING MARKET	NYSE
	ADDRESS	Dow Center Midland, MI 48674
	PHONE	800-232-2436
	CEO	Andrew N. Liveris
	WEB	*www.dow.com*

DESCRIPTION   Dow Chemical is the largest chemical company in the United States and the second largest in the world. It has a diversified portfolio in plastics, chemicals, and agricultural products and has consistently demonstrated an ability to apply science and technology to create new value-added products. It is one of the original developers of dendrimers—a nanoscale synthetic polymer that has a variety of applications.

REASONS TO BE BULLISH

► In 2005, Dow exchanged all of its dendrimers-related patents to Dendritic Nano-technologies (see page 208) for an equity stake in the company.

► The company has invested heavily in research and development with a significant component being directed to the nanosciences.

► Dow Corning (a joint venture of Dow and Corning) is working with the Institute of Soldiering Nanotechnologies—a state-of-the-art nanotechnology research center funded by the U.S. Army and the Massachusetts Institute of Technology—to develop next-generation materials for both soldiers and the commercial marketplace.

► It has a five-year agreement with Symyx (see pages 54–55) to help develop new nanomaterials and is partnering with Veeco (see page 59) to develop new nano-mechanical measurement instruments that will also help lead to the development of new nanomaterials.

► The company also has a good licensing agreement with the University of Texas to explore how nanoscale materials may help various drugs either be better absorbed into the body or more effectively target diseased cells.

continued

Dow Chemical continued

REASONS TO BE BEARISH

► The company currently has 60 percent of its facilities in the United States and could suffer if the dollar stays weak.

► It still faces litigation issues over asbestos, and if the cases are settled unfavorably, Dow's stock could suffer.

► The overall plastics industry is cyclical, and Dow will be adversely affected during any downturn.

WHAT TO WATCH FOR   Dow Chemical has successfully introduced RenaGel, a new drug for kidney failure patients, as well as other pharmaceutical products. Continue to look for new products and licensing opportunities coming out of its BioAqueous Solubilization Services division. (This division helps create more effective pharmaceutical products.) The announcement of any new drugs that treat cancer will be a positive sign. The company is also working on the development of a technology called Polymer Light Emitting Diode—which may lead to the development of a brighter, more flexible flat panel display that is durable and uses less energy. Lastly, Dow investors should monitor Dendritic Nanotechnologies' progress because it will benefit from that company's success.

CONCLUSION   Outlook is bullish. Dow Chemical is a well-run, diversified company with a strong international presence and a healthy approach to research and development. Its relationship with Dow Corning and its move into the life sciences sector suggests it will continue to diversify into new, higher-margin areas. The company should be part of any investor's portfolio.

DD	COMPANY	DuPont
	SYMBOL	DD
	TRADING MARKET	NYSE
	ADDRESS	1007 Market Street Wilmington, DE 19898
	PHONE	800-441-7515
	CEO	Charles O. Holliday
	WEB	*www.dupont.com*

DESCRIPTION DuPont is one of the world's largest chemical, materials, and energy companies and has businesses dedicated to electronics, communications, performance materials, coatings, agriculture, and nutrition. Nanotechnology plays a direct role in the following subdivisions: OLED displays, imaging technologies, titanium technologies, and personal protection materials.

REASONS TO BE BULLISH

► DuPont has a very solid R & D budget (approximately $1.2 billion) with a substantial portion of it dedicated toward nanotechnology as well as a strong portfolio of intellectual property.

► It is partnering with some of the most exciting energy-related nanotechnology companies (Nanosys and Konarka) in the development of flexible solar cells and Air Products and Chemical in the production of new nanomaterials.

► It is also partnering with—and investing heavily in—the U.S. Army's and the Massachusetts Institute of Technology's joint program at the Institute for Soldier Nanotechnologies where it is working to not only make Kevlar stronger and lighter, but to also clean water and make it germ resistant.

► The company has developed a proprietary method for producing carbon nanotubes of specific properties that could make them more practical for scores of applications.

► It is also a leading producer of titanium dioxide nanopowders, which have applications for cosmetics, environmental catalysts, and self-cleaning/self-sanitizing materials, and is well positioned to become a leader in the production of OLED (organic light emitting diode) technology.

continued

DuPont continued

REASONS TO BE BEARISH
► DuPont faces competition in the OLED market from Kodak and in the production of titanium dioxide nanomaterials from a host of smaller competitors.
► Its work with carbon nanotubes is still unproven.

WHAT TO WATCH FOR   If DuPont can continue its progress in the production of carbon nanotubes with predictable properties, it could open a host of new markets to DuPont. One huge market would be the production of biosensor/biosecurity devices that are capable of detecting even the faintest signs of contamination. The company is also working on something called "dynamic armor" that would create a bulletproof vest that hardens or softens as necessary.

CONCLUSION   Outlook is bullish. Given DuPont's exciting work with the Institute for Soldier Nanotechnologies, its relationship with Nanosys and Konarka, and its promising work with carbon nanotubes, investors would be wise to include the company in their portfolio.

EK	COMPANY	Eastman Kodak company
	SYMBOL	EK
	TRADING MARKET	NYSE
	ADDRESS	343 State Street Rochester, NY 00539
	PHONE	800-242-2424
	CEO	Daniel A. Carp
	WEB	*www.kodak.com*

DESCRIPTION   Eastman Kodak specializes in developing, manufacturing, and marketing professional, health, and other imaging products, including film, cameras, projectors, and photographic papers and chemicals. Its recent entry into the production of digital cameras—specifically its EasyShare LS633 digital camera, which features an OLED display designed with carbon-based molecules—has moved the company into the nanotechnology realm.

REASONS TO BE BULLISH

► Eastman Kodak is currently one of the world leaders in the production of OLED color screens and has a partnership with Sanyo Electric Ltd., which bodes well for future growth.

► Its investment in Nanosys (see page 230) may reap dividends and allow it to develop a leadership position in the area of flexible electronics.

► It is working to develop new nanoparticles that may help produce clearer instant photographs and reap dividends for the company as the market for digital camera printers continues to grow.

REASONS TO BE BEARISH

► The company is still suffering from the transition away from film-based cameras to digital cameras.

► In the area of OLED screens, Eastman Kodak faces competition in the form of Cambridge Display Technology whose manufacturing process may not only be more efficient but may also create a higher resolution and less expensive product.

continued

Eastman Kodak continued

WHAT TO WATCH FOR   If Eastman Kodak announces new partnerships for its OLED products, investors may want to reconsider the stock.

CONCLUSION   Outlook is bearish. At the present time, the downside risk of the company's fading film sales appears to outweigh the upside potential of its entry into the OLED market. Investors should approach the stock with caution.

》》   Nanotech in Europe

Infineon, formerly part of Siemens, has created the world's smallest carbon-nanotube-based transistor and is capable of competing with IBM, Hewlett-Packard, and Intel in this new emerging area.

Siemens, the European equivalent of General Electric and the continent's leading supplier of electrical parts and electronic equipment, is investing over $7 billion in research and development annually and is devoting a good portion of that money to nanotechnology—primarily through its Center for Functional Polymers.

The Henkel Group, which is much like 3M, is another big investor in nanotech and is applying it to its diverse product line. Hoechst AG and Bayer are also producing various nanoparticles and fullerenes, while Merck, the large pharmaceutical company, has established a relationship with C Sixty (now part of Carbon Nanotechnologies (see pages 83–84), GlaxoSmithKline, and Flamel (see page 173–74) to pursue various nanotechnology-enhanced pharmaceutical applications.

ELN	COMPANY	Elan Corporation
	SYMBOL	ELN
	TRADING MARKET	NYSE
	ADDRESS	Treasury Building Lower Grand Canal Street Dublin 2, Ireland
	PHONE	353-1-709-4000
	CEO	Kelly Martin
	WEB	*www.elan.com*

DESCRIPTION   Elan is a neuroscience-based technology company that is focused on discovering, developing, manufacturing, and marketing advanced therapies in neurology, autoimmune disease, and severe pain. It also owns Elan Pharma International that has successfully introduced its NanoCrytsal technology that is a drug delivery technology that takes advantages of the properties of nanoparticles to help improve the bioavailability of a number of drugs. (Note: At the present time, Elan Pharma constitutes only a small portion of Élan's overall sales.)

REASONS TO BE BULLISH

► In 2005, Roche announced it was licensing NanoCrystal to improve the bioavailability of it drugs, joining Merck, Bristol Myers Squibb, and Johnson & Johnson as companies already using the technology.

► Elan NanoCrystal technology is also currently being used by Johnson & Johnson in a drug to treat schizophrenia.

REASONS TO BE BEARISH

► Elan was not profitable in 2004, and its stock price plummeted in early 2005 after its Tysarbi drug (which is not related to nanotechnology and was thought to be a promising treatment for multiple sclerosis) was found to be responsible for killing one patient.

continued

Elan Corporation continued

WHAT TO WATCH FOR   If Elan Pharma meets Roche's development milestones for some of its drug candidates, it will be a positive sign (this holds true for its deals with the other pharmaceutical firms as well). In the next year, investors should also keep an eye on whether NanoCrystal's technology is approved by the FDA for the treatment of schizophrenia (which Johnson & Johnson is working on) and anticancer treatments (which EntreMed is seeking to develop).

CONCLUSION   Outlook is bullish. NanoCrystal has a host of applications and can be utilized by any number of major pharmaceutical firms. It is a very promising technology, however, until it becomes a larger portion of Elan's overall business, investors must assess Elan more for its pipeline of nonnanotechnology related drug candidates.

Fortune **500 companies**

EC	COMPANY	Engelhard Corporation
	SYMBOL	EC
	TRADING MARKET	NYSE
	ADDRESS	101 Wood Avenue Iselin, NJ 08830
	PHONE	732-205-5000
	CEO	Barry W. Perry
	WEB	*www.engelhard.com*

DESCRIPTION   Engelhard is a materials science company. It specializes in manipulating materials to alter their structure and surface characteristics. Many of its products are constructed at the molecular level and include a number of nanomaterials. It is the largest supplier of catalytic converters to the global auto industry.

REASONS TO BE BULLISH

► The company grew by 11 percent in 2004 and reported profits of nearly $200 million. In the first half of 2005, growth subsided slightly.

► By employing the unique characteristics of nanoscale materials, Engelhard can continue to refine and develop new materials not only for catalytic converters but also for a number of other markets, including petroleum refining, agriculture, coatings, cosmetics, and the plastics industries.

► One of its products, OxyClean, has been demonstrated to reduce nitrogen oxide emission by 45 percent (this would help petroleum refiners meet their standards).

REASONS TO BE BEARISH

► Although it has a diverse product line, a decline in automobile sales could hurt overall profits.

► Engelhard's environmental control technologies benefit from strict environmental standards and regulations. To the degree those standards are relaxed, the company may suffer from reduced product demand.

► It faces tough competition from Asian competitors such as Catalysts & Chemical Industries Co. and China Petroleum and Chemical Corp., as well as start-ups like Oxonica.

continued

Engelhard Corporation continued

WHAT TO WATCH FOR　Small start-ups like Oxonica and Catalytic Solutions, who are also working on new nanoscale catalysts, could disrupt Engelhard's market and its profits. Longer term, a transition away from the internal combustion engine to hybrid or fuel cell vehicles could evaporate its market for catalytic converters.

CONCLUSION　Outlook is bullish. Engelhard's diversity of markets, together with its understanding of the practical benefits of nanoscale materials, suggests it will have no problem continuing to find new applications that improve the bottom line of its customer base. Investors looking for a solid investment in the nanomaterials sector should consider Engelhard.

GE	COMPANY	General Electric
	SYMBOL	GE
	TRADING MARKET	NYSE
	ADDRESS	3135 Easton Turnpike Fairfield, CT 06828
	PHONE	203-373-2211
	CEO	Jeffrey Immelt
	WEB	*www.ge.com*

DESCRIPTION General Electric is the world's largest company (market capitalization of $380 billion) with eleven separate operating segments. GE manufactures everything from jet engines, gas turbines, and medical imaging systems to home appliances and water filtration products. In addition, it owns the NBC/Universal broadcasting system and has consumer and commercial lending and equipment leasing operations. It is, however, its Global Research Center that makes it a true nanotechnology company. In 2004, CEO Jeff Immelt, invested more than $5 billion in R & D and named nanotechnology one of GE's three priority areas.

REASONS TO BE BULLISH

► In 2004, GE's revenues increased 12 percent to $108 billion, and profits increased 2 percent to $11.2 billion. The positive trend continued in the first quarter of 2005.

► Its Global Research Center, which also has campuses in India and China, is working on nanoparticles to improve its medical imaging business and nanocomposites to make jet engines and turbines stronger, more temperature resistant, and quieter, and it is also producing hybrid nanomaterials for use in next-generation solar cells and fuel cells.

► Its work with carbon nanotubes and other nanomaterials could lead to new markets in the plastics and electronics industries.

continued

General Electric continued

REASONS TO BE BEARISH

► The sheer size of the company makes it difficult for GE to deliver extraordinary returns for the average investor.

► It now derives half of its earnings from its financial subsidiaries. While neither a positive nor a negative, may overshadow advances made possible by emerging technologies like nanotechnology.

WHAT TO WATCH FOR   If GE can develop breakthroughs for better medical imaging agents that can identify—and potentially treat—cancer and other diseases more quickly or if the company develops low-cost, high-efficiency photovoltaics, it would be wise to reconsider an investment.

CONCLUSION   Outlook is neutral. GE will likely continue to grow and is a solid investment, but investors would be better served holding onto the stock until it demonstrates more growth in the nonfinancial areas such as health care and energy.

Fortune 500 companies

HPQ		
	COMPANY	Hewlett-Packard
	SYMBOL	HPQ
	TRADING MARKET	NYSE
	ADDRESS	3000 Hanover Street Palo Alto, CA 94304-1185
	PHONE	650-857-1501
	CEO	Mark Hurd
	WEB	*www.hp.com*

DESCRIPTION   Hewlett-Packard is one of the world's largest information technology companies and has as its primary markets: printers, servers, and personal computer notebooks and data storage systems. In 1995, company founder, David Packard, created the Quantum Science Research Lab to pursue fundamental research in the areas of physical science—particularly the fabrication of nanoscale-scale structures and molecular electronics—that many believe will be the foundation of information technologies of the future.

REASONS TO BE BULLISH

► With an annual R & D budget of nearly $4 billion, HP is one of the world's leading inventors of new products and new technologies.

► In 2005, the company successfully demonstrated a "crossbar latch"—a molecular scale technology that behaves just like a transistor but is much smaller and easier to make, suggesting it may someday replace the transistor and lead to memory devices with a bit density ten times greater than today's silicon memory chips.

► Its relationship with Molecular Imprints (see page 68) suggests it is also scaling down to the sub-45 nanometer region for nearer-term applications.

► It has created (along with Oregon State University) an entirely new class of materials that could be used to make transparent transistors that are inexpensive, stable, and environmentally friendly. (Such devices could be incorporated in a variety of new uses including cheap gas sensors, automobile windshields that transmit visual information, and perhaps even "smart" clothing.)

continued

Hewlett-Packard continued

► HP has a deep portfolio of intellectual property and a strong scientific and techni-cal team.
► The company has the money and technical talent to pursue long-range, alterna-tive technologies such as atomic storage resolution.

REASONS TO BE BEARISH

► HP's existing businesses in imaging and printing systems, personal computers, storage, and servers have long been under pressure from the likes of IBM, Intel, and Dell.
► Although HP has increased its long-term investments in nanotechnology research and development, the company still lags behind the amount of money and human resources IBM is investing in the field.
► IBM, Intel, and a host of other private start-ups are all working on related or competing technologies.

WHAT TO WATCH FOR   If HP is successful in the large-scale manufacturing of devices that combine logic and memory, it could create some separation from itself and IBM and Intel. Longer term, if it can continue to perfect its molecular electronics, investors should consider adding shares.

CONCLUSION   Outlook is bullish. HP's work in molecular electronics is a calculated risk, but it appears to be paying off. If successful, the technology could eventually prove to be better, faster, and significantly cheaper than today's conventional methods for fabricating integrated circuits. For this reason alone, it is prudent to maintain at least a modest investment in HP.

Fortune 500 companies

HIT	COMPANY	Hitachi
	SYMBOL	HIT
	TRADING MARKET	NYSE
	ADDRESS	6, Kanda-Surugadai 4-chome Chiyoda-ku, 101-8010 Tokyo, Japan
	PHONE	650-244-7900
	CEO	Etsuhito Shoyama
	WEB	*www.hitachi.com*

DESCRIPTION   Hitachi, an $80 billion conglomerate, is one of Japan's largest companies and manufactures everything from auto parts and electrical generators to medical devices, computer hard drives, and telecommunications devices. In 2001, the company merged three separate units to create the Hitachi High-Technologies Corporation for the specific purpose of developing, manufacturing, and marketing new products and services in the emerging field of nanotechnology.

REASONS TO BE BULLISH

► Hitachi has announced its intention to begin commercializing low-cost "nanostamp" technology for medical applications, which it claims will drive the price of a DNA chip down to around $10 to $12. (Given the company's large size and deep pockets, it may be able to beat its competitors to the market with such a product.)

► In 2004, Hitachi increased its nanotechnology budget to more than $100 million.

REASONS TO BE BEARISH

► The company is overly dependent on the Japanese economy, which accounts for 70 percent of sales.

► In the nanolithography area, it faces tough competition from promising nanotech start-ups such as Molecular Imprints (see page 68) and Obducat (see pages 51–52)—both of whom have relationships with some of Hitachi's larger competitors.

► In the area of DNA chips, the company faces competition from Affymetrix (see page 118–119).

continued

Hitachi continued

WHAT TO WATCH FOR   Hitachi has developed some promising nanoimprint lithography, but it is difficult to compare it with that of its competitors. If, however, the technology is either sold to or licensed to other companies, Hitachi's stock may be worth reconsidering.

CONCLUSION   Outlook is bearish. Hitachi has a large R & D budget (nearly $4 billion), but it does not appear to be positioned as the leader in any of the most promising new emerging industries. Investors would be better served looking elsewhere for an investment.

IBM	COMPANY	IBM
	SYMBOL	IBM
	TRADING MARKET	NYSE
	ADDRESS	1133 Westchester Avenue White Plains, NY 10604
	PHONE	800-IBM-4YOU
	CEO	Sam Palmisano
	WEB	*www.ibm.com*

DESCRIPTION   IBM is the biggest computer equipment vendor and information technology services provider in the world and has a history of creating, developing, and manufacturing the industry's most advanced information technologies, including computer systems, software, networking systems, storage devices, and microelectronics. The company also has a number of nanotechnology-specific projects in nanoelectronics, nanomaterials, bionanotechnology, and material characterization.

REASONS TO BE BULLISH

► In 2004, revenues increased 8 percent to $96 billion, and profits increased to $8.65 billion—an 11 percent increase over the previous year. In the first half of 2005, revenues continued to increase but not at as fast a pace as in 2004.

► With its state-of-the-art research labs all around the world, a $6 billion annual R & D budget, a world-class team of scientists, and more than 700 nanotechnology-related patents, IBM is well positioned to continue its leadership position in the decade ahead.

► The company has a history of using technology to expand existing markets and create new markets; for instance, its work in nanomaterials could help create smaller and cheaper RFID tags and distributed nanosensors.

► Its Millipede technology (expected to reach the market in 2006 or 2007) could capture a large share of the data storage market or have unique applications in a variety of products.

continued

IBM continued

► IBM is a leader in the development and use of carbon nanotubes and has created the first nanotube-based logic circuit—which might enable dramatically improved circuits and data storage devices—and is also exploring carbon nanotubes' ability to absorb and emit light, which could lead to advancements in fiber-optic technology, including radically faster bandwidths.

► It has achieved breakthroughs in nanoscale magnetic resonance imaging (MRI) that could lead to the creation of microscopes capable of making three-dimensional images of a molecule. (Such an advancement would lead to a much better understanding of biological structures.)

► IBM has created a self-assembling template that could possibly be used to construct computer circuits down to the 20 nm range; it has developed a new type of strained germanium (which is more conductive than silicon) and can potentially build small, more powerful computer chips (down to the 20 nanometer range).

► The company is partnering with Stanford University to create the Spintronic Science and Applications Center that might establish IBM as a leader in spintronics—using the spin of electrons to store data—and significantly increase the density of hard disks and double the bandwidth capacity of existing wires.

REASONS TO BE BEARISH
► As a $90 billion company, IBM is unlikely to achieve substantial annual growth;
► Hewlett-Packard, Intel, Motorola, and a significant number of private start-ups are working on competing nanotechnologies. In the first quarter of 2005, the company shed nearly 20 percent of its value due to slowing sales.

WHAT TO WATCH FOR   In the near term, watch for an announcement that its Millipede technology will be incorporated into products; midterm, look for IBM to create carbon tube circuits; and longer-term, expect its nanotechnology-related research to lead it more into the life sciences sector.

CONCLUSION   Outlook is bullish. IBM has been—and will be—a leader in the development of new technologies. Given the size of its research and development budget and its strong scientific team, the company should be a part of every investor's portfolio, and investors should not be put off by short-term fluctuations in revenues.

INTC	COMPANY	Intel Corporation
	SYMBOL	INTC
	TRADING MARKET	NYSE
	ADDRESS	2200 Mission College Boulevard Santa Clara, CA 95053
	PHONE	408-765-8080
	CEO	Paul S. Otellini
	WEB	*www.intel.com*

DESCRIPTION   Intel is the world's largest manufacturer of personal computer semiconductor components, including microprocessors (e.g., Pentium and Centrino), memory chips, graphic chips, and various network and communication products. It currently controls approximately 80 percent of the global market for these products. In late 2004, Intel outlined its future product roadmap for the next decade, and it was clear that nanomaterials and nanotechnology would play a dominant role in its future.

REASONS TO BE BULLISH

► In 2005, Intel spent close to $5 billion on R & D (an amount close to the annual revenues of its largest competitor Advanced Micro Devices), and a good portion of that money was dedicated to exploring carbon nanotubes and nanowires to help the company meet its internal goals of producing chips with features in the 10 nanometer range by 2011.

► It already has existing relationships with two very promising nanotechnology-related start-ups—Zyvex, with whom it is investigating whether carbon nanotubes can be dispersed into polymers and help serve as a thermal interface (a development that would help keep the chips cool as they get ever smaller and hotter), and with Nanosys to work on memory-related nanotechnologies.

► Intel Capital, the company's venture capital arm, has deep pockets and has demonstrated a willingness to invest in promising or disruptive nanotechnology-related companies.

continued

Intel Corporation continued

REASONS TO BE BEARISH

► With close to 85 percent of its sales coming from personal computers, the company must rely heavily on replacement cycles that, for such a mature market, may be difficult to replace.

► Infineon, IBM, and AMD are all strong competitors, and if their nanotechnology-related developments make it to market first, it could take significant market share away from Intel. In the memory market, Hewlett-Packard and ZettaCore (see page 239) are strong competitors.

WHAT TO WATCH FOR   Intel, in addition to employing nanotechnology to enable better products, is also exploring new uses for its products in two potentially big markets. The first is the work the company is doing with sensor networks, and the second relates to its work analyzing and identifying proteins to detect the presence of cancer or other diseases.

CONCLUSION   Outlook is bullish. Intel appears very well situated for the long run. It has consistently invested in cutting-edge technologies and has a compelling vision of the future—a future that embraces a pervasive computing environment. Moreover, it has demonstrated that it is not afraid to take risks, and it does not punish its managers or leaders if those risks don't pay off. Because of its aggressiveness, strong R&D budget, venture capital arm, and well-known marketing prowess, the company should be included as part of any investor's portfolio.

Fortune **500 companies**

MOT		
	COMPANY	Motorola
	SYMBOL	MOT
	TRADING MARKET	NYSE
	ADDRESS	1303 East Algonquin Road Schaumburg, IL 60196
	PHONE	847-576-6873 (Investor Relations)
	CEO	Ed Zander
	WEB	*www.motorola.com*

DESCRIPTION   Motorola is the third largest manufacturer of mobile phones in the world and has a significant manufacturing presence in the wireless, broadband, and cable segments. (Although in late 2004, it spun off its semiconductor unit into a separate company, Freescale Semiconductor (FSL).) Motorola's Physical Sciences Research Lab focuses much of its efforts on advancing nanotechnology in the areas of material science, nonvolatile semiconductor memory, molecular self-assembly, and nanoelectronics.

REASONS TO BE BULLISH

► In 2004, the company increased revenues by 27 percent and reported a profit of nearly $2.2 billion (up from $900 million the year before). The positive trend continued in the first half of 2005.

► Motorola is doing research in the area of carbon nanotubes that could lead to the creation of low-cost "nano emissive flat panel displays" as soon as 2006. In May of 2005, the company unveiled its first prototype in this area.

► It is working on developing a conductive "nano-Velcro" packaging interconnect that would enable the manufacture and assembly of electronics with solder or adhesives, and its work in the area of nanomaterials may help produce better batteries, antennas, sensors, and coatings.

► As an investor in Molecular Imprints (see page 68), Motorola could also be well positioned to develop next-generation flash memory chips.

continued

Motorola continued

REASONS TO BE BEARISH

► In its key market (mobile phones), it has lost ground to Nokia and Samsung, and the entire sector faces severe downward pricing pressure.

► None of its nanotechnology research has yet made its way into commercial products that are competitive in the marketplace.

WHAT TO WATCH FOR   If Motorola does come to the market with a low-cost flat panel display, investors should consider investing in the company. Also, the company's efforts to commercialize printable organic field-effect transistors bear watching because it could usher in an era of cheap, flexible electronics and lead to such things as product packaging with animated features and updateable newspapers.

CONCLUSION   Outlook is bearish. Motorola is a good company and understands how nanotechnology can be used to keep the company competitive in the long run. However, until it can actually get some of its nanotechnology products to the market, investors should treat the stock with caution.

NIPNY	COMPANY	NEC Corporation
	SYMBOL	NIPNY
	TRADING MARKET	NYSE (traded as ADR)
	ADDRESS	7-1, Shiba 5-Chome Minato-Ku, Tokyo, 108-8001 Japan
	PHONE	81-3-3454-1111
	CEO	Hajime Sasaki
	WEB	*www.nec.com*

DESCRIPTION   NEC (formerly known as Nippon Electric Company) is a leading Japanese technology company with three separate product lines: information technology, network solutions, and electronic devices. It manufactures a variety of personal computers, semiconductors, display panels, and broadband and mobile communication devices. NEC researchers discovered carbon nanotubes in 1991, and the company is a leader in the emerging field of quantum computing.

REASONS TO BE BULLISH

▶ In 1991, a NEC researcher, Sumio Iijima, discovered carbon nanotubes. His research has positioned NEC as a leader in how the material might be employed in a variety of devices, and his discovery might also be protected by patent law, which could require many of the companies presently manufacturing carbon nanotubes to seek a licensing agreement with NEC.

▶ The company has demonstrated carbon nanotubes as field-effect transistors and is working to develop nanobiochip technology.

▶ It is a world leader in the development of "carbon nanohorns" that have demonstrated significant promise in increasing the efficiency of fuel cells. (The company employed the carbon nanotubes in fuel cells for use in cell phones and laptop computers in 2005, and longer term, the materials might help make automobile fuel cells a viable alternative to the internal combustion engine.)

▶ NEC's excellent relationships with the Japan Science and Technology Agency and the National Institute for Material Sciences will help the company stay atop the latest developments in the field of nanotechnology.

continued

NEC Corporation continued

REASONS TO BE BEARISH

► NEC is highly dependent on the Japanese economy. The country's continued slow growth or a recession will have a negative impact on NEC.

► The company employs aggressive accounting techniques and has a complex corporate structure that makes assessing the company and its stock difficult.

► In the area of employing carbon nanotubes as transistors, NEC faces stiff competition from IBM and Infineon.

► Its work in the fuel cell market, while promising, may not meet with widespread consumer acceptance. Consumers may not warm to the idea of having to refill— instead of recharge—their cell phones or laptops, and the price of carbon nanohorns is currently prohibitive for a number of applications.

WHAT TO WATCH FOR   If NEC's carbon nanohorns are adopted for the automobile fuel cell market, investors would be wise to increase their holdings in the company stock. Similarly, if NEC continues to demonstrate significant progress in the development of quantum–cryptographic technology (its devices can reportedly securely transmit data over 150 kilometers—currently a world record), it would be an even more bullish sign.

CONCLUSION   Outlook is bullish. NEC has suffered from eight years of slow growth, and its corporate and accounting practices, as well as its overexposure to the Japanese economy, make it a risky stock. However, the company's research and development in the area of carbon nanotubes, carbon nanohorns, and quantum computing suggest that the prudent nanotechnology investor should have a small stake in the company.

PHG	COMPANY	Royal Philips Electronics
	SYMBOL	PHG
	TRADING MARKET	NYSE
	ADDRESS	Breitner Center Amstelplein 2 Amsterdam, 1096 The Netherlands
	PHONE	877-248-4237
	CEO	Gerard Leisterlee (Chairman of the Management Board)
	WEB	*www.philips.com*

DESCRIPTION   Philips Electronics is Europe's largest manufacturer of electronic and consumer products, including CD and DVD players, television, and medical imaging systems. The company also designs semiconductors. It has dedicated extensive resources to nanotechnology research and development.

REASONS TO BE BULLISH

► Philips employs more than 2,000 researchers and scientists at its research laboratories (and devotes approximately 8 percent of its annual sales to research), and it is currently working on developing silicon nanowires for next-generation electronics devices, nanowires made of indium phosphide for OLEDs, and quantum dots for medical imaging devices.

► The company is working with FEI (see pages 43–44) to develop a new type of electron microscope using carbon nanotubes.

► It is a principle partner in IMEC's—one of the world's leading independent semiconductor research organizations—sub-45 nm CMOS research program and will help in its efforts to develop integrated circuits down to the 10 nanometer range.

REASONS TO BE BEARISH

► As a large conglomerate, the company has been inefficiently managed, and in key sectors, particularly semiconductors and medical imaging devices, the company faces stiff competition.

continued

Royal Philips Electronics continued

WHAT TO WATCH FOR   The company has done a lot of work in the field of electronic paper. Currently, the devices are still difficult to read, and they have not yet been able to demonstrate a full color device. If Philips overcomes these obstacles, it may open a host of new product lines and would warrant further consideration as an investment.

CONCLUSION   Outlook is bearish. Philips offers a diverse product line but is currently only a leader in the field of lighting. In the semiconductor and medical imaging businesses, it trails other companies such as IBM, Intel, and General Electric. At the present time, it does not appear well positioned to deliver above-market returns.

Summary

Almost every major corporation in the world is involved in nanotechnology to some degree. Some, like Hewlett-Packard, are more public about their efforts, while others prefer to take a quieter approach. Investors should look for companies that have a pragmatic, multipronged approach when it comes to nanotechnology initiatives: (1) a short-term strategy that stresses nanotechnology's application in improving existing products, (2) a midterm strategy that focuses on partnering with potentially "disruptive" nanotechnology start-ups who could transform their existing businesses, and (3) a longer-term emphasis on investing in the research and development projects to ensure their products will still be cutting edge ten, fifteen, and twenty years down the road.

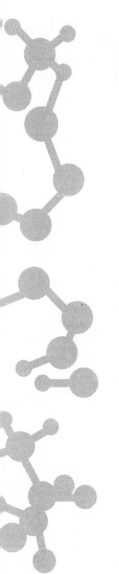

"The biggest shift we've seen in the last year is nanotechnology moving out of the lab and onto the production line. Nanotech startups are beginning to make money, with revenue ranges between $10 million to $20 million for those at the top of the ranks. They're partnering with established corporations to develop products in a pattern similar to biotech: Ten of the 30 corporations in the Dow Jones Industrial Average have announced nanotech partnerships."

—Matthew Nordan, Vice President of Research at Lux Research

Chapter 6

Small Can Be Beautiful: Small to Midsize Publicly Traded Nanotechnology Companies

The companies profiled in Chapters 3, 4, and 5 are tame compared to what follows. The publicly traded small to midsize cap companies (under $1 billion in market capitalization) profiled in the following pages can be likened to the Wild West of nanotech investing. Unlike the *Fortune* 500 companies, who are only tangentially involved in nanotech, these companies are heavily invested in nanotechnology. Whereas the equipment and nanomaterials suppliers offered a relatively modest risk-to-reward ratio, these companies because they are supplying (or are trying to supply) real nanotechnology solutions to real problems offer a higher risk-to-reward ratio.

If past history is any guide and it is likely that it is, these stocks are not for the faint-of-heart. It will be common to see some of their prices jump or fall by as much as 10, 20, or even 30 percent in a single day. There will be a host of reasons for this. All of which the individual investor must be aware. To begin, because many are involved in the pharmaceutical business, their products are regulated by the Food and Drug Administration, and approval of their drug or product can send their stock soaring. Alternatively, a failing grade can send it spiraling downward. The FDA's approval of American Pharmaceutical Products nanoparticulate drug in early 2005 is a perfect example. The day after Abraxane was approved, the company's stock soared 50 percent.

Many of these smaller companies are also dependent on partnerships and relationships with larger companies for their long-term success. For instance, Flamel, another company applying nanoparticles for drug delivery, suffered a severe setback in late 2004 when Bristol Myers Squibb canceled its partnership on Basulin (a nanoparticle drug designed to treat diabetes); the announcement that Cypress (a large semiconductor manufacturer) was getting out of the line of business in which it intended to use NVE Corporation's proprietary nanotechnology to manufacture a next-generation memory chip similarly sent its stock southward. The bottom line is that these smaller companies are often just the tail on a much larger dog, and as the dog wags its tail, so goes the small company.

Some of the companies are also difficult to value because their product is so new and the market they are going after is not yet matured. For instance, Harris & Harris is a publicly traded venture capital company specializing in nanotechnology. Many of the nanotech start-ups it has invested in have great potential. Unfortunately, some of the companies don't have any commercial products yet, let alone customers. The risk of pursuing a big market with an as yet untested technology is what makes investing in these stocks so risky—and potentially rewarding.

Another common risk associated with smaller company stocks is that, in their early stages, they are often overly reliant on government contracts and don't have an established commercial customer base. Others are valued according to their intellectual property—which may or may not prove to be valuable. Of course, all are subject to the competitive pressures of the marketplace. Often bigger, better funded companies are pursuing the same markets, while many of the private "disruptive" nanotechnology companies covered in Chapter 8 are trying to attack the same markets—only from a completely different angle.

Lastly, investors need to understand that a few companies listed in this chapter are traded as pink sheet and bulletin board stocks. These stocks are very risky because they do not receive the same level of oversight as stocks traded on NASDAQ or the New York Stock Exchange. They are not required to disclose as much financial information, and therefore, their status is more difficult to assess. They also tend to be thinly traded, which means that their stock price can fluctuate wildly on the sale of just a few shares. Investors need to beware of glowing press releases issued by some of these companies because they are sometimes purposely designed to artificially heighten interest in the company's stock in an attempt to drive the stock price up.

This isn't to say that all the companies traded on these smaller markets are bad—two Australian companies—Starpharma and pSivida—are two exceptions. In general, however, investors should tread cautiously before investing in companies that are traded on alternative markets.

Having said all of this, there are some really solid companies making real products and going after large established markets that offer the more risk-tolerant investor some attractive investments. Here is a list of those smaller nanotechnology companies whose stocks are publicly traded today.

AVNA	COMPANY	Advance Nanotech
	SYMBOL	AVNA
	TRADING MARKET	Over-the-counter
	ADDRESS	712 5th Avenue, 19th Floor New York, NY 10019
	PHONE	646-723-8962
	CEO	Magnus Gittins
	WEB	*www.advancenanotech.com*

DESCRIPTION Advance Nanotech specializes in the acquisition, incubation, and commercialization of promising nanotechnologies. The company currently has eighteen portfolio nanotechnology companies—seven in electronics, six in biopharma, and five in materials.

REASONS TO BE BULLISH

► The company completed a $20 million financing in early 2005 and has limited resources to pursue its business model.

► It has assembled an impressive corporate advisory board who—if they are actually involved—may provide a good deal of assistance to the portfolio companies.

► The company continues to work at establishing promising academic partnerships, such as the one it has with the University of Cambridge.

REASONS TO BE BEARISH

► The company's business model is untested and it has yet to achieve any commercial marketplace success.

► It has no revenues.

► In every field—electronics, biopharma, and materials—it faces stronger and better established competitors.

WHAT TO WATCH FOR If Advance Nanotech can really help its portfolio companies gain access to funding, quality management, and scientific and technical advice, and thus facilitate the rapid commercialization of its portfolio companies technology, it has the potential to be profitable.

continued

Advance Nanotech continued

CONCLUSION   Outlook is bearish. The company's business model is too unproven and its financial resources are too small to give it much chance of a success. Investors are encouraged to wait to see if it can deliver even one success. If it does, investors would be encouraged to reconsider an investment at that time. Presently, the company is just too much of an unknown to recommend.

APPX		
	COMPANY	American Pharmaceutical Partners
	SYMBOL	APPX
	TRADING MARKET	NASDAQ
	ADDRESS	1101 Perimeter Drive, Suite 300 East Schaumburg, IL 60173-5837
	PHONE	847-969-2700
	CEO	Alan L. Heller
	WEB	*www.appdrugs.com*

DESCRIPTION   American Pharmaceutical Partners is a subsidiary of American Bio-Science, Inc. and is a drug company specializing in developing, manufacturing, and marketing injectable, generic pharmaceutical products. In 2005, the FDA approved Abraxane, a proprietary nanoparticle oncology drug that binds to naturally occurring proteins and more effectively delivers dosages, while also reducing toxicity (meaning fewer side effects).

REASONS TO BE BULLISH

► The approval of Abraxane by the FDA served as a significant validation of the company's proprietary nanoparticle technology and opens the way for the company to gain a significant share of the $1 billion market for Taxol—which treats breast and ovarian cancer.

► APPX's revenues increased from $11 million in the first quarter of 2004 to more than $24 million in the first quarter of 2005.

► The company's generic drug business (it currently has over 135 injectable drugs) gives the company a solid base, and its ability to transfer its nanoparticle technology to other drugs promises additional future growth.

REASONS TO BE BEARISH

► Even with the approval of Abraxane, APPX will still face competition from other companies seeking to develop similar treatments. For instance, Bristol Myers Squibb could reformulate Taxol or a smaller private start-up, such as SoluBest, could introduce a product with superior qualities.

continued

American Pharmaceutical Partners continued

► Abraxane may also still give patients some problematic side effects.
► In the past, the Securities and Exchange Commission (SEC) has investigated the company for potentially misleading investors about the progress of some of its drugs.

WHAT TO WATCH FOR   The company's success rests on its ability to transfer its technology to new drugs. Investors need to monitor if it can get new treatments into the FDA pipeline. Investors should also monitor the company's P/E ratio. If it gets too far out of line with the standards for the pharmaceutical industry, it would be cause for concern.

CONCLUSION   Outlook is bullish. Given the approval of Abraxane, its solid generic drug business, healthy good gross margins (over 50 percent), and strong cash position ($60 million), the company appears well positioned for future growth.

ARWR	COMPANY	Arrowhead Research Corporation
	SYMBOL	ARWR
	TRADING MARKET	NASDAQ
	ADDRESS	1118 East Green Street Pasadena, CA 91106
	PHONE	626-792-5549
	CEO	R. Bruce Stewart
	WEB	*www.arrowres.com*

DESCRIPTION   Arrowhead Research Corporation is not a nanotechnology company in the sense that it actually produces a product. Rather, the company funds nanotechnology research at universities (primarily Caltech) in return for the right to commercialize the technology and/or license the resulting intellectual property. To date, the company consists of three subsidiaries: Aonex Technologies, which is developing semiconductor nanomaterials; Insert Therapeutics, which is developing nanoscale drug delivery systems; and Nanotechnica, which is reportedly working on the production of a variety of nanoscale devices. In late 2004, Merrill Lynch added the company to its Nanotechnology Index.

REASONS TO BE BULLISH
▶ Arrowhead's CEO, Bruce Stewart, has achieved past success in building businesses with a similar intellectual property-based model, and it is possible that the company's portfolio contains some valuable patents.

REASONS TO BE BEARISH
▶ Arrowhead is currently generating very limited revenues and a high cash burn rate (about $2 million per quarter).
▶ Aonex, its semiconductor nanomaterials company, faces a great deal of competition from much larger and better funded competitors.

continued

Arrowhead Research Corporation continued

► Insert Therapeutics has not yet entered Phase I clinical trials, suggesting that even if its devices are demonstrated to be effective they are still years away from profitability.

► Nanotechnica's business model lacks specification and is reportedly close to closing down.

WHAT TO WATCH FOR   If Insert Therapeutics' drug candidate advances to Phase II clinical trials, the stock might be worth considering at that time.

CONCLUSION   Outlook is bearish. Investors should not be misled into believing that just because Merrill Lynch added this stock to its Nanotechnology Index that it is a prudent investment. Investors are cautioned to stay away until Arrowhead has actually licensed some of its technology and that technology has been used in a commercial product.

Small to Midsize Publicly Traded Nanotechnology Companies

BDSI	COMPANY	BioDelivery Sciences International
	SYMBOL	BDSI
	TRADING MARKET	NASDAQ
	ADDRESS	4 Bruce Street Newark, NJ 07103
	PHONE	973-972-0015
	CEO	Francis O'Donnell
	WEB	*www.biodeliverysciences.com*

DESCRIPTION   BioDelivery Sciences is a biotechnology company that is developing and attempting to commercialize Bioral, a nanocrystalline drug delivery technology.

REASONS TO BE BULLISH
► Bioral has a host of potential applications, including the effective delivery of small molecules and proteins.
► Bioral is also made out of food safe ingredients, suggesting it has the possibility of being used to add nutrients and antioxidants in food products to improve their health efficacy.

REASONS TO BE BEARISH
► BioDelivery had minimal revenues for in the first quarter of 2005 and a net loss of approximately $1 million.
► It has no major pharmaceutical or food companies as customers to date.

WHAT TO WATCH FOR   The company expects to launch its first commercial product in the first quarter of fiscal year 2006. If this timeline is met and if the product is received favorably in the commercial marketplace, investors should revisit the stock. BioDelivery is also working on BEMA Fentanyl, currently in Phase I clinical trials, which promises to help relieve cancer pain. Investors should watch if the product advances to Phase II clinical trials. If it does, it would be a bullish sign.

CONCLUSION   Outlook is bearish. BioDelivery has very promising technology, but until it actually lines up customers, investors should treat the company with caution.

BIPH.OB	COMPANY	Biophan Technologies, Inc.
	SYMBOL	BIPH.OB
	TRADING MARKET	Over-the-counter
	ADDRESS	150 Lucius Gordon Drive, Suite 215 West Henrietta, NY 14586
	PHONE	585-214-2441
	CEO	Michael Weiner
	WEB	*www.biophan.com*

DESCRIPTION   Biophan is developing a proprietary thin film nanomagnetic particle coating to make biomedical devices MRI (Magnetic Resonance Imaging) safe and image compatible.

REASONS TO BE BULLISH

► Many of today's medical devices—particularly pacemakers and neurostimulators—cannot be used in MRI equipment because the magnetic fields cause the devices to heat up and cause tissue damage. Biophan's technology reportedly alleviates this problem.

► Biophan has a relationship with Boston Scientific (BSX), is partnering with NASA to create a biothermal battery that converts the body's natural heat into usable electrical energy (such batteries could extend the life of cardiac pacemakers and require fewer surgical replacements), and has a strong intellectual property portfolio.

REASONS TO BE BEARISH

► Biophan has no revenues and lost $4 million in 2004. In the first quarter of 2005, it also reported no revenues.

► No companies yet employ Biophan's proprietary technology.

continued

Biophan Technologies, Inc. continued

WHAT TO WATCH FOR   The employment of Biophan's technology in pacemakers represents the largest market, followed by neurostimulators. If it can successfully enter either market, it would serve as a validation of its business model, and investors may want to revisit the company. Other devices such as catheters, guide wires, and endoscopes offer promising but smaller markets. The company has also launched a subsidiary, Nanolution LLC, to develop drug delivery applications. This bears watching, but due to the need to obtain FDA approval, any commercially viable product is likely years away.

CONCLUSION   Outlook is bearish. Biophan's intellectual property is promising. Its relationship with Boston Scientific is a positive, but until medical device manufacturers actually start licensing—and employing—its technology, Biophan remains a risk.

BPA	COMPANY	BioSante Pharmaceuticals
	SYMBOL	BPA
	TRADING MARKET	AMEX
	ADDRESS	111 Barclay Boulevard Lincolnshire, IL 60069
	PHONE	847-478-0500
	CEO	Stephen M. Simes
	WEB	*www.biosantepharma.com*

DESCRIPTION   BioSante Pharmaceuticals is developing a pipeline of hormone therapy products to treat both men and women. It is also developing a calcium phosphate nanoparticle drug delivery system for the oral delivery of insulin and other vaccines.

REASONS TO BE BULLISH

► Biosante's calcium phosphate nanotechnology has demonstrated some success in animal trials and its Bio-E-Gel (which treats hot flashes) is nearing the completion of phase III clinical trials.

► The company has entered into agreements with the U.S. Army, Navy, and Defense Department to develop noninjected vaccines to protect against malaria and a variety of bioterror agents, including anthrax, ricin, staph, and the bubonic plague.

REASONS TO BE BEARISH

► The company only had revenues of $23,000 in 2004 and losses of nearly $8 million. The company reported an additional loss of $2.8 million for the first quarter of 2005.

► BioSante's primary business—developing hormone therapy products for male and female sexual dysfunction—has encountered difficulty in gaining FDA approval.

WHAT TO WATCH FOR   FDA approval of Bio-E-Gel would give BioSante a huge push and likely boost the company's revenues. Disapproval, however, will likely cause the company to float additional shares to generate the necessary cash to stay afloat.

CONCLUSION   Outlook is bearish. Although BioSante is listed on Merrill Lynch's nanotechnology index, the company's nanoparticle drug delivery system represents its only involvement in nanotechnology. The technology bears watching, but until it reaches Phase II clinical trials, the downside risk of owning BioSante stock is too great.

CALP	COMPANY	Caliper Life Sciences, Inc.
	SYMBOL	CALP
	TRADING MARKET	NASDAQ
	ADDRESS	68 Elm Street Hopkinton, MA 01748
	PHONE	508-435-9500
	CEO	Kevin Hrusovsky
	WEB	*www.caliperls.com*

DESCRIPTION   Caliper Life Sciences is a microfluidics company using advanced liquid handling and lab chip technologies for drug discovery and molecular diagnostics.

REASONS TO BE BULLISH

► The company's revenues have been consistently increasing over the past four years, and it has almost $50 million in cash on hand and little debt.

► Caliper's acquisition of Zymark in 2003, its long-standing relationship with Agilent, along with a new partnership with Affymetrix to automate target preparation could take its genomic research to a larger—and possibly industrial—scale and position it well for future growth.

REASONS TO BE BEARISH

► Caliper reported a $25 million loss in 2004.

► The lab-on-a-chip field is very competitive, and longer term, private start-ups such as Fluidigm and NanoString, could disrupt Caliper's technology.

WHAT TO WATCH FOR   If Affymetrix extends its relationship with Caliper for genomics research or if the company moves into the molecular diagnostic market, longer-term growth is likely. Moreover, if its technology proves to be a catalyst for the discovery of cancer therapeutics, it will be a bullish signal.

CONCLUSION   Outlook is bullish. For the short term (2006), Caliper appears nicely positioned to help pharmaceutical companies cut their drug discovery cycles. Its throughput rates are higher than the industry standard, and its data is of superior quality than that of its peers.

Small to Midsize Publicly Traded Nanotechnology Companies

CPHD	COMPANY	Cepheid, Inc.
	SYMBOL	CPHD
	TRADING MARKET	NASDAQ
	ADDRESS	904 Caribbean Drive Sunnyvale, CA 94089
	PHONE	408-541-4191
	CEO	John Bishop
	WEB	*www.cepheid.com*

DESCRIPTION   Cepheid develops, manufactures, and markets microfluidic systems that integrate, automate, and accelerate biological testing, including DNA analysis and the detection of various biothreats.

REASONS TO BE BULLISH

► Cepheid's revenues have been growing consistently, and in 2004, they more than doubled over the previous year.

► The company is partnering with Northrop Grumman on a $175 million contract with the U.S. Post Office to supply its GeneXpert system to more than 290 post offices to help rapidly identify a variety of infectious organisms, including anthrax.

► It has a multimillion dollar deal with the U.S. Army to develop a system for detecting biothreats and other pathogens.

REASONS TO BE BEARISH

► In spite of increasing sales, the company still has not generated a profit.

► The company faces stiff competition from companies such as Affymetrix and Nanogen, and it is overly reliant on contracts with the U.S. government—specifically the U.S. Post Office and the U.S. Army.

WHAT TO WATCH FOR   If Cepheid can move into the clinical diagnostics or the cancer detection markets, it will significantly improve sales. Also, in the event of a major biological attack, the demand for Cepheid's products would likely grow.

CONCLUSION   Outlook is bearish. The company bears watching, but until it can move beyond being primarily a government contractor and broaden its sales to the commercial marketplace, investors should treat it with caution.

CBMX	COMPANY	CombiMatrix Corporation
	SYMBOL	CBMX
	TRADING MARKET	NASDAQ
	ADDRESS	6500 Harbour Heights Parkway, Suite 301 Mukilteo, WA 98275
	PHONE	425-493-2000
	CEO	Amit Kumar, PhD
	WEB	*www.combimatrix.com*

DESCRIPTION   CombiMatrix is the life sciences business of Acacia Research Corporation and primarily develops biochips for DNA analysis in the emerging field of genomics and proteomics. It is also partnering with another company, Nanomaterials Discovery, to use its biochips to help discover new nanomaterials.

REASONS TO BE BULLISH

► CombiMatrix's revenues appear to be growing fast, and it is profitable.

► Its relationship with Roche Diagnostics along with a sizeable grant from the Defense Department to develop microarrays for detecting biological threats make the prospects for future growth promising.

REASONS TO BE BEARISH

► The company is overly reliant on government contracts.

► In its primary area of biochips, it faces competition from a large number of more well-established firms, like Affymetrix, and in the field of developing new nanomaterials, it is in direct competition with Symyx and Nanostellar.

WHAT TO WATCH FOR   If the company's new nanomaterials business actually develops some new materials that are licensed by major manufacturers, it would be a validation of its technology and would warrant a second look at the company's stock.

CONCLUSION   Outlook is bearish. CombiMatrix's technology is promising, but it faces too much competition in the area of DNA analysis. Until the company can demonstrate a consistent ability to generate a profit, investors should tread cautiously.

Small to Midsize Publicly Traded Nanotechnology Companies

CTMI		
	COMPANY	CTI Molecular Imaging, Inc.
	SYMBOL	CTMI
	TRADING MARKET	NASDAQ
	ADDRESS	810 Innovation Drive Knoxville, TN 37932-2571
	PHONE	865-218-2000
	CEO	Ronald Nutt, Ph.D.
	WEB	*www.ctimi.com*

DESCRIPTION   CTI Molecular Imaging, Inc. is a manufacturer of position emission tomography (PET) imaging equipment. It is a leading producer of molecular biomarkers that are essential for imaging changes in a person's body and can help signal cancer, disease, or neurological disorders.

REASONS TO BE BULLISH
► CTI is profitable, and revenues grew by over 10 percent in 2004.
► The company is one of the largest suppliers of PET equipment, as well as the largest supplier of biomarkers to the PET market.
► The number of producers using PET is rapidly growing, and the fact that its technology is reportedly superior at detecting the presence of various diseases at a much earlier stage than competing technologies promise suggests that future revenue growth is very likely.
► The company is partnering with both Siemens and Hitachi to distribute its products globally.

REASONS TO BE BEARISH
► The biomarkers essential for determining the proteins that characterize Alzheimer's disease (a very lucrative area) are not yet on the market.
► New emerging technologies—such as functional magnetic resonance imaging—could preclude the use of external markers and render CTI's products obsolete.
► In late 2004, the Centers for Medicare and Medicaid Services reduced the reimbursement for PETs, which could adversely affect sales.

continued

CTI Molecular Imaging, Inc. continued

WHAT TO WATCH FOR   Investors will want to consider increasing their stake in CTI if it successfully introduces a biomarker for Alzheimer's disease.

CONCLUSION   Outlook is bullish. Given the overall growth of PET and CTI's relatively strong position in the field, the company appears to be a solid investment. Furthermore, the threat of functional MRIs taking significant market share away from CTI is a distant threat.

EMFP.OB	COMPANY	Emergency Filtration Products
	SYMBOL	EMFP.OB
	TRADING MARKET	Over-the-counter
	ADDRESS	175 Cassia Way, Suite A115 Henderson, NV 89014
	PHONE	702-558-5164
	CEO	Douglas Beplate
	WEB	*www.emergencyfiltration.com*

DESCRIPTION   Emergency Filtration Products develops specialty air filtration technology for removing infectious bacteria and viruses in airflow systems. Most of the company's products do not deal with nanotechnology, but it does market and sell a product called NanoMask, which reportedly utilizes nanoparticles to isolate and destroy viral and bacterial contaminants.

REASONS TO BE BULLISH
▶ The company has sold products to the U.S. military in the past, and in 2004, it entered into a distribution agreement with Itochu Techno Chemical to distribute some of the company's products in Asia.

REASONS TO BE BEARISH
▶ Sales for the first quarter of 2005 were down 92 percent over the same period last year and totaled only $170,000.
▶ The company's stock has been subject to such wild fluctuations that it prompted officials at the company to take the unusual step of asking investors to take physical possession of their stock certificates.

WHAT TO WATCH FOR   In the event of a disease-borne outbreak, such as SARS, investors may want to monitor demand for the company's product. If it increases, the company may be able to stabilize and grow its operations.

CONCLUSION   Outlook is bearish. Emergency Filtration is currently too small, too risky, and its relationship to nanotechnology too tenuous to recommend.

ENEI.OB	COMPANY	Ener1
	SYMBOL	ENEI.OB
	TRADING MARKET	Over-the-counter
	ADDRESS	500 West Cypress Creek Road, Suite 100 Fort Lauderdale, FL 33309
	PHONE	954-556-4020
	CEO	Kevin Fitzgerald
	WEB	*www.ener1.com*

DESCRIPTION   Ener1 develops and markets new technologies including lithium batteries, fuel cell components, and nanomaterials through its subsidiaries: EnerDel, EnerFuel, and NanoEner.

REASONS TO BE BULLISH

► In 2005, the company raised $14 million in a private placement, thus ensuring it has some money to develop its technologies.

REASONS TO BE BEARISH

► Through March of 2005, the company only generated $33,000 in revenues.

► In every area it is pursuing, it faces stiff competition from more experienced and better financed companies.

WHAT TO WATCH FOR   Investors should be very leary of any and all promising press releases.

CONCLUSION   Outlook is very bearish. Lithium batteries, fuel cell technology, and nanomaterials are all very competitive areas. A company with such stiff competition and only minimal revenues does not warrant a market cap of $100 million. Investors should stay away from this stock.

FLML	COMPANY	Flamel Technologies
	SYMBOL	FLML
	TRADING MARKET	NASDAQ
	ADDRESS	33 Avenue du Dr. Georges Levy 69693 Venissieux Cedex France
	PHONE	33.0.472.783.434
	CEO	Stephen H. Willard
	WEB	*www.flamel.com*

DESCRIPTION   Flamel Technologies is one of the true leaders in developing and manufacturing innovative, nanoscale drug delivery systems. Its Medusa technology, which is optimized for the controlled-release delivery of drugs, promises to more effectively deliver drugs and do so with fewer side effects (this is because the nanoparticles are nontoxic and thus require no solvents).

REASONS TO BE BULLISH

► In 2004, Flamel significantly increased revenues and became profitable and continued the trend in the first quarter of 2005.

► If the company can perfect its Medusa technology, it could lead to licensing opportunities and create a healthy revenue stream for the foreseeable future.

► The company's Micropump technology (which is not nanotechnology) could be used to successfully reformulate a number of blockbuster drugs (it presently has agreements with Biovail and GlaxoSmithKline).

REASONS TO BE BEARISH

► The licensing of Basulin, a product designed to treat diabetes, has twice been terminated by major pharmaceutical companies—suggesting that there is some problem with the drug (for instance, it may not release drug molecules at the prescribed rate). It has also raised questions about the company's credibility.

► Its other promising drug candidates—Interleukin-2 and Interferon alpha—are both in Phase I/II clinical trials and may not prove successful.

continued

Flamel Technologies continued

WHAT TO WATCH FOR   Partnerships and FDA approval of future drug candidates are the keys to Flamel's success. More cautious investors should wait until the company's various drugs have successfully completed Phase II clinical trials before considering an investment.

CONCLUSION   Outlook is bullish. Flamel has a very strong intellectual property position with over fifty patents, and its technology is promising. Investors with a high tolerance for risk should consider investing in the stock. The market for therapeutic protein drugs is so large that even if only one of the company's drugs meets with success, the stock is likely to appreciate considerably. American Pharmaceutical Product's success with Abraxane speaks to the viability of Flamel's business model.

Small to Midsize Publicly Traded Nanotechnology Companies

TINY	COMPANY	Harris & Harris Group Inc.
	SYMBOL	TINY
	TRADING MARKET	NASDAQ
	ADDRESS	111 West 57th Street New York, NY 10019
	PHONE	212-582-0900
	CEO	Charles Harris
	WEB	*www.hhgp.com*

DESCRIPTION   Harris & Harris Group is a publicly traded venture capital company that specializes in making investments in nanotechnology and microelectromechanical systems (MEMS) companies. To date, it has equity stakes in eleven nanotechnology companies. In many ways, Harris & Harris represents the only way for individual investors to gain an equity stake in some of the most disruptive nanotechnology companies.

REASONS TO BE BULLISH

► Harris & Harris has positions in the following companies: Molecular Imprints, NanoMix NanoOpto, NanoPharma, Nanosys, Nanotechnologies, Inc., Nantero, NeoPhotonics, Optiva, Starfire Systems, and Zia Lasers all of which hold great promise.

► The company's management and technical expertise is highly regarded, and two of its nanotechnology investments (Nanophase and NanoGram) have already been sold at a healthy profit.

REASONS TO BE BEARISH

► The stock is very hard to price. The current book value per share (also Net Asset Value (NAV)) is currently around $4.50 per share, meaning that investors are paying a healthy premium for the privilege of investing in its high-risk ventures at its 2005 trading range of between $10 and $14.

continued

Harris & Harris Group Inc. continued

WHAT TO WATCH FOR   As Chapter 7 explains, almost all of the companies Harris & Harris have taken an equity stake in have the potential for significant return-on-investment. Molecular Imprints, Nantero, and Nanosys are especially attractive. Investors should track developments in those companies, and if any of the companies appear on the verge of major developments, an investment in Harris & Harris is a smart way to make a play on that progress.

CONCLUSION   Outlook is bullish. Harris & Harris is an excellent company and has a solid approach toward investing in nanotechnology. As a venture capital company, however, investors must understand that more often than not, the companies Harris & Harris invests in will not become profitable. (The few that do, however, are expected to more than offset those losses.) As a rule of thumb, investors should consider the stock fairly priced if it is around double its Net Asset Value.

HDWR	COMPANY	Headwaters, Inc.
	SYMBOL	HDWR
	TRADING MARKET	NASDAQ
	ADDRESS	10653 South River Front Parkway, Suite 300 South Jordan, UT 84095-3529
	PHONE	801-984-9400
	CEO	Kirk A. Benson
	WEB	*www.hdwtrs.com*

DESCRIPTION   Headwaters develops and commercializes technologies to enhance the value of coal, gas, oil, and other natural resources. It also owns Headwaters Technology Group, Inc., a company that is developing nanocatalysts to convert coal and heavy oils into higher yield, environmentally-friendly liquid fuels.

REASONS TO BE BULLISH

► Headwaters is profitable and has grown from sales of $45 million in 2001 to an estimated $554 million in 2004, while substantially lowering its debt.

► The company's nanocatalyst technology is not currently priced into the stock. It can significantly increase the value chain of coal, and its ability to upgrade heavy oil (of which Canada has a huge reserve) by making it anywhere between 10 to 20 percent lighter could be very lucrative.

REASONS TO BE BEARISH

► The company is dependent on Section 29 of the U.S. Tax Code for a substantial tax credit (it essentially makes coal dust profitable). This code is up for renewal in 2007, and if it is not renewed, the company's profits will take a severe hit. (In 2004, the tax credit accounted for nearly 40 percent of its profits.)

► Its investment in nanotechnology, which is required for coal liquefaction, is relatively small (less than $5 million) and may not be enough to bring about this transition.

► Also if the oil price drops, the demand for upgrading heavy oil or coal liquefaction could decrease.

continued

Headwaters, Inc. continued

WHAT TO WATCH FOR

► If the company's nanocatalyst technology proves successful in upgrading heavy oil, it will be a bullish sign. An even more bullish sign will be the successful conversion of coal to a liquid fuel. The latter would allow the company to take one ton of coal (at a cost of $12) and convert it into four barrels of oil at a price of $30 to $60 per barrel.

CONCLUSION   Outlook is bullish. Headwaters had a huge run-up in 2004 (from $19 to $34), and future growth may now become more difficult. However, its newer nanotechnologies offer an attractive play for aggressive investors looking for a partial nanotechnology play in the energy sector.

IMMC	COMPANY	Immunicon Corporation
	SYMBOL	IMMC
	TRADING MARKET	NASDAQ
	ADDRESS	3401 Masons Mill Road, Suite 100 Huntingdon Valley, PA 19006
	PHONE	215-830-0777
	CEO	Edward L. Erickson
	WEB	*www.immunicon.com*

DESCRIPTION   Immunicon Corporation develops a variety of equipment and other technologies—including magnetic nanoparticles (also called ferrofluids)—that are used by researchers, diagnostic technicians, and drug companies to identify, detect, and count the number of tumor cells present in a small blood sample. Immunicon's first product helps monitor how breast cancer patients react to a variety of different treatments. Given the fact that over two million women in the United States have been diagnosed with breast cancer and 46,000 die annually from it, the market is significant. Immunicon was one of the first nanotech companies to go public in April 2004.

REASONS TO BE BULLISH
► Immunicon's magnetic nanoparticles are capable of detecting and attaching themselves to individual cancer cells and promise to be able to detect cancer very early (which will improve the odds of patient survival).
► The nanoparticles can also be used to help determine the efficacy of new cancer drugs—making the product very useful to major pharmaceutical companies, as well as letting patients know if certain treatments are working.
► In 2004, the company signed an extension with Pfizer to use its products to help determine the effectiveness of certain experimental therapies.
► It already has established relationships with Johnson & Johnson (Veridex), Pfizer, and Invitrogen—all three could lead to significant royalty income. It has a very strong intellectual property position with over ninety patents to its credit.

continued

Immunicon Corporation continued

REASONS TO BE BEARISH

► The stock is down from its original IPO price and has yet to generate a profit.

► Its exclusive license with Veridex for its cancer cell analysis products could limit future sales.

WHAT TO WATCH FOR   The company is reportedly working on cell kits to help detect prostate, lung, ovarian, and a wide variety of other cancers, as well as cardiovascular and infectious diseases. If it is successful—especially in the area of cardiovascular disease—it would be worth considering accumulating additional shares.

CONCLUSION   Outlook is bullish. Although Immunicon's stock price (at the time of this publication) was below its IPO price, its technology has been proven effective, and it shows great promise in detecting a variety of other cancer cells. While it is heavily reliant on its partnership with Johnson & Johnson's subsidiary Veridex, there is a good chance that it will develop additional strategic alliances (or sign licensing agreements) with other major pharmaceutical companies.

KOPN	COMPANY	Kopin Corporation
	SYMBOL	KOPN
	TRADING MARKET	NASDAQ
	ADDRESS	200 John Hancock Road Taunton, MA 02780
	PHONE	508-824-6696
	CEO	John C.C. Fan
	WEB	*www.kopin.com*

DESCRIPTION   Kopin is a developer and manufacturer of telecommunication and digital imaging technologies, including heterojunction bipolar transistor (HBT) wafers and liquid crystal display technologies. The company's displays employ what it calls NanoPocket technology—a novel method of confining the production of light away from defects—to create thinner, clearer, and more energy-efficient devices.

REASONS TO BE BULLISH

► Kopin's revenues increased 24 percent in 2004 to nearly $70 million, and it is currently supplying displays to 30 percent of the world camcorder market.

► Its customers include Samsung, JVC, Matsushita (Panasonic), Boeing, and Nokia, and the Defense Department is now using its technology in a host of devices.

REASONS TO BE BEARISH

► Kopin still lost $7 million in 2004 and revenues were off the first part of 2005.

► The display market is very competitive, and Kopin faces tough competition from Hitachi and a host of start-ups.

WHAT TO WATCH FOR   Kopin has stated that it believes it can make its LED technology three times brighter and ten times cheaper. If these aggressive goals can be met, Kopin should be a leader in the field. Another bullish sign would be if the automotive industry embraced LED lighting and began making it standard on all new automobiles.

CONCLUSION   Outlook is bullish. Kopin's growth, strong intellectual property position (it has over 200 patents), and its established relationship with major manufacturers, together with the inroads it has made into the U.S. military, suggest that it is well positioned for future growth.

LMRA	COMPANY	Lumera Corporation
	SYMBOL	LMRA
	TRADING MARKET	NASDAQ
	ADDRESS	19910 N. Creek Parkway Bothell, WA 98011-3008
	PHONE	425-415-6900
	CEO	Thomas D. Mino
	WEB	*www.lumera.com*

DESCRIPTION   Lumera develops polymer nanomaterials as well as products based on those materials for use in the wireless, optical, and life sciences sectors. The company went public in July 2004 and is included on Merrill Lynch's Nanotechnology Index.

REASONS TO BE BULLISH

► Lumera boasts the financial backing of Cisco and Intel Corp.

► It holds the rights to a great deal of intellectual property from the University of Washington.

REASONS TO BE BEARISH

► The company reported losses of over $6 million on sales of just over $1 million for 2004, and in the first quarter of 2005 revenues continued to decrease.

► It currently has no commercial products, and while it is distributing "samples" to major manufacturers, it has been unable to secure any lasting deals. (In fact, in late 2004, the company dropped its first product—a wireless antenna—with little reason.)

► In almost every area of production, it faces much stronger and better financed competition.

WHAT TO WATCH FOR   If Cisco or Intel were to turn from investor to customer, it might warrant taking a second look at this stock.

CONCLUSION   Outlook is bearish. Until Lumera actually sells some real products to major manufacturers, the stock's downside appears to outweigh its upside potential. Longer term, the company is not investing enough in research and development to be competitive in its stated areas of emphasis.

Small to Midsize Publicly Traded Nanotechnology Companies

XDSL.OB	COMPANY	mPhase Technologies, Inc.
	SYMBOL	XDSL.OB
	TRADING MARKET	Over-the-counter
	ADDRESS	587 Connecticut Avenue
		Norwalk, CT 06854
	PHONE	203-838-2741
	CEO	Ronald A. Durando
	WEB	*www.mphase.com*

DESCRIPTION   mPhase is a developer of broadband communications products, including digital subscriber line (DSL) products for telecommunication service providers. It is also a development stage technology company and has, with the help of the New Jersey Nanotechnology Consortium (Bell Labs), developed a "nanograss battery technology"—which reportedly allows a battery's electrodes to last significantly longer.

REASON TO BE BULLISH

► mPhase's technology, if successful, would produce a battery that is potentially smaller, less expensive, and more effective than today's conventional batteries.

REASONS TO BE BEARISH

► For the first quarter of 2005 (the last period for which sales figures were available), the company generated only $564,000 in sales.

► A number of other companies, including Gillette, Power Paper, and Cap-XX, are investing heavily in improved battery technology, and mPhase faces stiff competition.

WHAT TO WATCH FOR   If major battery manufacturers begin licensing mPhase's technology or it is incorporated into real products (such as medical device implants that require a battery or radio frequency identification tags), investors will want to reconsider the stock. Another bullish sign would be if the U.S. military was to employ mPhase's technology in its sensor networks.

CONCLUSION   Outlook is bearish. mPhase's small annual sales (less than $1.5 million for 2004) and the competitive nature of the battery market suggest that the downside risk of this stock far outweighs its upside potential.

NNBP.OB	COMPANY	Nanobac Life Sciences
	SYMBOL	NNBP.OB
	TRADING MARKET	Over-the-counter
	ADDRESS	2727 W. Martin Luther King Boulevard, Suite 850 Tampa, FL 33607
	PHONE	813-264-2241
	CEO	John Stanton
	WEB	*www.nanobaclifesciences.com*

DESCRIPTION   Nanobac is a biolife science company that is attempting to develop diagnostic tests and treatment for nanobacteria that are purported to play a role in coronary artery calcification (CAC). CAC, in turn, could be associated with coronary artery disease—one of the leading causes of death in the United States.

REASONS TO BE BULLISH
- In 2004, the Mayo Clinic conducted some research that suggested that nanobacteria may exist.
- Nanobac has sponsored research that demonstrates a correlation between the presence of antibodies to nanobacteria and CAC.

REASONS TO BE BEARISH
- In spite of the Mayo Clinic research, many in the medical profession dispute the existence of nanobacteria.
- Nanobac's tests and products have not yet received FDA approval.
- Its over-the-counter stock status and management structure make it a risk.

WHAT TO WATCH FOR   If nanobacteria are found to exist, the company's first move should be to develop a cost-effective diagnostic test. Investors should then track FDA developments regarding the production of medications that can safely and effectively treat nanobacteria.

CONCLUSION   Outlook is bearish. Until further independent studies have been done establishing the existence of nanobacteria and then more clearly establish a link between their presence and heart disease, investors should simply monitor this company's developments.

Small to Midsize Publicly Traded Nanotechnology Companies

NGEN		
	COMPANY	Nanogen, Inc.
	SYMBOL	NGEN
	TRADING MARKET	NASDAQ
	ADDRESS	10398 Pacific Center Court San Diego, CA 92121
	PHONE	858-410-4600
	CEO	Howard Birndorf
	WEB	*www.nanogen.com*

DESCRIPTION   Nanogen develops, commercializes, and supplies molecular diagnostic products and tests to help understand how genes function and to better understand their correlation between genetic variation and disease. In spite of Nanogen's name and the fact that it has a series of products revolving around its Nanochip technology—an electronic microarray—the company's only direct relationship with nanotechnology is through some of its intellectual property.

REASONS TO BE BULLISH
► The company has recently received some patents in the area of nanofabrication (self-assembly) that may lead into some promising new areas such as molecular electronics.
► Its recent merger with Epoch Biosciences is expected to nearly double its revenues.

REASONS TO BE BEARISH
► Nanogen has yet to post a profit, and it has a high price-to-sales ratio.
► Its lead competitor in the area of molecular diagnostics, Affymetrix, is not only profitable, but it is also much better funded and has a much stronger sales and distribution network.

WHAT TO WATCH FOR   If Nanogen is able to license some of its nanofabrication intellectual property to major manufacturers, the company would be worth revisiting.

CONCLUSION   Outlook is bearish. In the field of microarrays, Affymetrix is far stronger and a better investment than Nanogen. Moreover, the company's only direct ties to nanotechnology are in areas where a commercially viable product is still off in the distant future.

Small to Midsize Publicly Traded Nanotechnology Companies

NVAX	COMPANY	Novavax, Inc.
	SYMBOL	NVAX
	TRADING MARKET	NASDAQ
	ADDRESS	508 Lapp Road Malvern, PA 19355
	PHONE	484-913-1200
	CEO	Nelson Sims
	WEB	*www.novavax.com*

DESCRIPTION   Novavax is a biopharmaceutical company engaged in the research, development, and commercialization of drug delivery products. It has developed a micellar "nanoparticle" technology (the particles are larger than 100 nanometers), which is an oil, water, and lipid topical emulsion product used for delivering hormone treatments.

REASONS TO BE BULLISH
▶ Novavax's nanoparticle technology, Estrasorb, was approved by the FDA in 2003, and the company has entered into a partnership with King Pharmaceuticals in an attempt to enter the $2 billion estrogen therapy market.
▶ It also has created, Androsorb, a topical testosterone emulsion that is currently in clinical trials.

REASONS TO BE BEARISH
▶ Novavax is not profitable yet, and it faces stiff competition from a number of larger, better financed pharmaceutical companies.

WHAT TO WATCH FOR   If major pharmaceutical companies begin licensing its micellar nanoparticle technology for the topical delivery of other drugs, investors should revisit the stock.

CONCLUSION   Outlook is bearish. Estrasorb, while effective, has not yet gained a significant portion of the estrogen therapy market, and until it does, the stock is too pricey.

NVEC		
	COMPANY	NVE Corporation
	SYMBOL	NVEC
	TRADING MARKET	NASDAQ Small Cap
	ADDRESS	11409 Valley View Road Eden Prairie, MN 55344
	PHONE	952-829-9217
	CEO	Daniel Baker
	WEB	*www.nve.com*

DESCRIPTION   NVE manufactures sensors and couplers used for acquiring and transmitting data in automated factories. It is, however, the company's intellectual property in the area of spintronics—the utilization of an electron's spin, rather than its charge, to acquire, store, and transmit information—that makes it a nanotechnology company. (The technology promises to store a lot more information and use less energy.) If NVE's patents in this area hold up, it could allow the company to license its IP to a variety of semiconductor manufacturers and garner significant royalties.

REASON TO BE BULLISH

► The MRAM market could reach an estimated $50 billion by 2010. If NVE's IP is an essential component of MRAM technology, the company could generate significant royalty revenues.

REASONS TO BE BEARISH

► NVE has already licensed its technology to Motorola and Cypress, but in 2005, Cypress announced it was getting out of the business, which has placed a big cloud over its viability.

► The company is currently profitable. But its sales to this point have nothing to do with nanotechnology. Its market capitalization is quite high for a company with such limited revenues.

continued

NVE Corporation continued

► Although Motorola has licensed its technology, its spin-off, Freescale Semiconductor, may not be beholden to that agreement.
► IBM, Hewlett-Packard, Intel, and Samsung are also investing in MRAM technology. Given NVE's small size, it may not be able to compete against these giants.

WHAT TO WATCH FOR   If Freescale announces it is using NVE's IP, it will be another bullish sign. Investors will also want to pay attention to any royalty agreement the company signs with future licensors of the technology.

CONCLUSION   Outlook is bearish. NVE holds some valuable IP, but IBM and others are likely to enter the commercial marketplace with competing technologies.

VITA	COMPANY	Orthovita, Inc.
	SYMBOL	VITA
	TRADING MARKET	NASDAQ
	ADDRESS	45 Great Valley Parkway Malvern, PA 19355
	PHONE	610-640-1775
	CEO	Anthony Koblish
	WEB	*www.orthovita.com*

DESCRIPTION   Orthovita, Inc. is a biomaterials company that has developed proprietary technologies—including nanoparticles—to facilitate bone repair.

REASONS TO BE BULLISH

► Orthovita's lead product, VITOSS, is a highly porous, resorbable nanoparticulate that acts as a scaffold for cell seeding to help damaged bone heal.

► VITOSS is an attractive alternative to using bone material from cadavers to repair serious spinal injuries.

► In July 2004, the company received $25 million in financing from Cohesion Technologies to aid the development of new products and expand the company's sales force.

► Revenues increased from $10 million in 2003 to $17.6 million in 2004, and it has almost $30 million in cash.

REASONS TO BE BEARISH

► Orthovita has not yet generated a profit, and its products have not yet captured the attention of orthopedic surgeons.

WHAT TO WATCH FOR   As attractive as the market for surgical repair is, an even larger market is the treatment of osteoporosis. The development of any technology that treats this disease should cause investors to give the company another look. (Note: FDA approval would likely push the commercial development of such a product far out to the future.)

CONCLUSION   Outlook is bearish. Orthovita has promising technology, but until VITOSS is used in a significantly larger portion of spine, hip, and knee surgical procedures and begins running a profit, investors should treat the stock with caution.

PSD	COMPANY	psivida Limited
	SYMBOL	PSD (PSD.AX) or PSVD.PK
	TRADING MARKET	Australian Stock Exchange (Also traded as an ADR)
	ADDRESS	Level 12, BGC Centre 28 The Esplanade Perth, WA 6000 Australia
	PHONE	61.8.9226.5099
	CEO	Roger Brimblecombe
	WEB	*www.psivida.com*

DESCRIPTION   pSivida is an Australian-based biotechnology company committed to the biomedical applications of nanotechnology, specifically the development and commercialization of BioSilicon—biocompatible and biodegradable nanostructured porous silicon—that has multiple applications including controlled drug delivery, brachytherapy, and tissue engineering.

REASONS TO BE BULLISH

► pSivida's BrachySil technology (a radioactive biochip that "locks" on cancer cells and releases predetermined doses of radioactive molecules to kill the cell) has been demonstrated to be effective in reducing the number of malignant cells in patients with inoperable liver cancer. (By reducing the tumor, the device can help prolong a patient's life long enough to receive a potentially lifesaving transplant.)

► The company's BioSilicon, which is silicon manufactured with nanopores that can be loaded with drugs, genes, proteins, and other therapeutics or vaccines, can also be tailored to release its contents over a defined period of time. (Such a technology could free patients from having to take regular dosages of pills.)

► It has agreements with Itochu, one of Japan's largest life science companies, to help identify, develop, and commercialize BioSilicon in Asia and has an agreement with NanoHorizons (see page 221) to examine whether its technology can be employed in thin films to help facilitate tissue engineering.

► pSivida also has a strong family of intellectual property with over twenty patents and enough cash on hand ($30 million) to operate until at least 2007.

continued

pSivida Limited continued

REASONS TO BE BEARISH

► The company is not currently profitable. (It lost approximately $3 million in 2004.)
► Its BioSilicon and BrachySil technology face competition from a host of other drug delivery devices and cancer-fighting drugs.
► The company's success is dependent on regulatory approval for its products.

WHAT TO WATCH FOR   If pSivida's BrachySil gains FDA approval or if the company is successful in developing a "smart" drug delivery device (a device that could electronically release the right amount of a drug at the right location), investors should consider adding pSivida to their holdings.

CONCLUSION   Outlook is bullish. pSivida is a good investment. Although its two leading technologies have not yet received final approval from American, European, or Asian regulators, the technology looks very promising. Furthermore, the platform technology holds the potential to be used for a variety of other applications. In January 2005, it began trading on the U.S. market under the Symbol: PSVD.PK.

SKYE	COMPANY	SkyePharma PLC
	SYMBOL	SKYE
	TRADING MARKET	NASDAQ
	ADDRESS	105 Piccadilly London W1J 7NJ England
	PHONE	44.207.491.1777
	CEO	Michael Ashton
	WEB	*www.skyepharma.com*

DESCRIPTION   SkyePharma is a specialty pharmaceutical company developing nanoparticulate drug delivery technology to improve a variety of existing drug delivery technologies.

REASONS TO BE BULLISH

► Approximately 50 percent of all drug candidates are abandoned by pharmaceutical companies due to the body's inability to absorb the drug. SkyePharma's technology has the potential, by improving drug solubility, to remove this barrier, as well as open up a variety of new opportunities by producing precisely prescribed nanoparticles that can more effectively reach their targets.

► Due to the nanoparticles' small size, the drugs don't require any toxic solvents, thus reducing adverse side affects.

► The company already has agreements with Endo Pharmaceuticals, Merck, Quintiles, and AstraZenaca and expects three new drugs employing its technology to be launched by 2006.

► SkyePharma revenues increased by 26 percent in 2004, and it has a strong R & D team.

REASONS TO BE BEARISH

► The company lost $47 million in 2004 and faces a lot of competition from companies such as Eli Lilly, Elan, Flamel, Solubest, and American Pharmaceutical Products.

continued

SkyePharma PLC continued

WHAT TO WATCH FOR   The more agreements SkyePharma signs with the big drug companies, the better its stock will do. Similarly, if its technology is used to improve or reformulate blockbuster drugs, investors will want to consider increasing their holdings.

CONCLUSION   Outlook is bullish. SkyePharma's technology has the potential to help a number of major drug manufacturers by either making their drugs more effective, opening them up to new markets, or extending their patents through reformulation. This diversity of clients, products, and markets positions the company well for future growth.

Small to Midsize Publicly Traded Nanotechnology Companies

SPL.AX/ SPHRY. PK	COMPANY	Starpharma Pooled Development Ltd.
	SYMBOL	SPL.AX/SPHRY.PK
	TRADING MARKET	Australian Stock Exchange (Also traded as an ADR in the United States)
	ADDRESS	Level 6, Baker Heart Research Building Commercial Road Melbourne, Victoria Australia
	PHONE	61.3.9510.5955
	CEO	Dr. John Raff
	WEB	*www.starpharma.com*

DESCRIPTION   Starpharma is a biopharmaceutical company focused on the development and application of dendrimer nanotechnologies as drug delivery devices. Its lead product, VivaGel, has received clearance from the FDA for human clinical trials.

REASONS TO BE BULLISH

► VivaGel is a dendrimer-based topical microbicide gel that has been developed for women as a preventative treatment against the sexual transmission of HIV. The gel has been proven 100 percent effective in animal trials and has demonstrated early positive signs in human trials.

► Starpharma's technology will also be tested for effectiveness against other sexually transmitted diseases such as Herpes and Chlamydia. (An estimated 15 million women in the United States contract sexually transmitted diseases every year.)

► In early 2005, Starpharma agreed to give Dow Chemical a sizeable stake in Dendritic Nanotechnologies in exchange for the rights to all of its dendrimer-based intellectual property.

continued

Small to Midsize Publicly Traded Nanotechnology Companies

Starpharma Pooled Development Ltd. continued

► Starpharma still owns the largest share of Dendritic Nanotechnologies (DNT), which is one of the few companies to be accepted as a partner in the Institute of Soldiering Nanotechnologies. (It is likely that its dendrimers are being explored as a platform for detecting and treating various chemical and biological agents.) Furthermore, in mid-2005, DNT announced that it had developed a new class of dendrimer called the Priostar that was much easier and less expensive to manufacture than previous dendrimers.

► In 2004, a consortium lead by Starpharma was awarded $5 million from the National Institutes of Health to develop a second generation microbicide for the prevention of HIV and other STDs.

REASONS TO BE BEARISH

► The company is not profitable and has few revenues at the present time.

► NanoBio (see page 216) as well as other major pharmaceutical companies are also investing in similar or competing dendrimer-based technologies.

WHAT TO WATCH FOR   If a major pharmaceutical company licenses Starpharma's technology, investors should consider an increased investment. If the FDA does not approve VivaGel, it would be a severe blow to the company's future prospects. Investors should also be on the lookout for additional applications for dendrimer-based technology such as the delivery of nutraceuticals.

CONCLUSION   Outlook is bullish. Starpharma is a good investment for those investors with a stronger penchant for risk. VivaGel's effectiveness in animal studies, coupled with the company's large stake in Dendritic Nanotechnologies, provide great potential upside—especially in the area of treating various cancers.

WEDX	COMPANY	Westaim Corporation
	SYMBOL	WEDX
	TRADING MARKET	NASDAQ
	ADDRESS	144 4th Avenue SW, Suite 1010 Calgary, AB T2P 3 Canada
	PHONE	780-992-5231
	CEO	Barry Heck
	WEB	*www.westaim.com*

DESCRIPTION Westaim Corporation operates two very distinct subsidiaries. iFire Technology develops and commercializes flat panel displays (and is not involved with nanotechnology), and Nucryst manufacturers antimicrobial wound care products that utilize a nanocrystalline form of silver.

REASONS TO BE BULLISH

► Nucryst's Acticoat dressing is used in 100 of the 120 burn centers in North America.

► The company has successfully met two separate $5 million milestones from Smith & Nephew, and still has another $24 million in potential milestone payments.

► Nucryst is in Phase II clinical trials for a topical treatment of eczema, a skin disorder that affects approximately 20 million Americans and represents a significantly larger market than wound dressings.

► It has over $117 million in cash and no debt, and the company recently began a $7 million expansion of its production facility in Canada.

REASON TO BE BEARISH

► While Nucryst is profitable, Westaim Corporation is not, and its other subsidiary, iFire, is a small player in a very competitive market.

WHAT TO WATCH FOR If the FDA approves Nucryst's eczema product or if it develops other nanocrystalline products to treat arthritis or cancer, it will be a bullish sign.

CONCLUSION Outlook is neutral. If Nucryst was a stand-alone company, it would be a solid investment. It is Westaim's other subsidiary, iFire, which is troubling and makes the company's stock difficult to assess.

Summary

Investing in small and midsized nanotechnology companies is comparable to the Wild West of investing. The analogy is apt for many reasons, not the least of which is that it was often not the pioneer of the Wild West who profited but the settlers who followed afterward. The pioneers, as the old saying goes, were rewarded for their efforts with a lot of arrows in their back. The settlers, however, were allowed to stake their claims in relative peace.

Much the same thing will happen in the field of nanotechnology. Many of these smaller companies are the pioneers, and they are pushing the envelope. Some will succeed but many will fail.

Investors are encouraged to look for the following four items when assessing these companies. First, focus on management. Specifically, ask yourself if they have the experience or a proven track record of success? Second, determine if the company has superior technology; it may be the only thing that can guarantee their survival. Third, determine if the companies have a partnership with an established larger company. Such a relationship can provide some protection against unpredictable market factors. Finally, you should look at companies that are focusing on the most attractive and high-margin markets. Be wary of companies that are trying to be all things to all people. For the foreseeable future, as the world grows larger, older, and more interconnected, there will be a continued need for energy, health care, and information technology-related products. Nanotechnology companies that are focused on these areas and can develop products to meet the growing needs are likely to be rewarded.

"It's hard to think of an industry that isn't likely to to be disrupted by nanotechnology."
—David Bishop, vice president of Nanotechnology Research at Bell Labs

"[Nanotechnology] will be bigger than the Internet and more far-reaching. It will create vast new wealth. It will destroy a lot of old wealth. And it will shake up just about every business on the planet."
—Forbes Wolfe Nanotech Report

Chapter 7

The Waves of Changes: The Disruptors

Joel Barker in his book, *Paradigms: The Business of Discovering the Future*, begins with a short story about the Swiss watchmaking business. He noted that in the late 1960s the Swiss controlled 65 percent of the world market (and 80 percent of its profits) in high-quality watches. A decade later, however, they controlled only 10 percent and were forced to lay off 50,000 of the 62,000 people they employed in the watchmaking business. The reason: quartz technology.

What is interesting is that Swiss scientists actually invented the technology, but their brethren in the watch industry refused to accept it because they couldn't grasp how it applied to their business. In short, they felt they were in the business of making mechanical watches, and they could not conceive of making an electronic watch. Unfortunately for them, Seiko and Texas Instruments could, and they wasted little time taking advantage of the opportunity. Such is the power of a disruptive technology.

History is replete with similar examples—everything from the transistor to the Internet. Technological change is one of the primary forces—if not the primary—for change in today's society. This pace is only growing more explosive, and nanotechnology is going to be one of its primary catalysts.

As the companies profiled in this chapter will give testament, nanotechnology is poised to enable solar cells capable of being "painted on" walls, memory chips of almost unfathomable density, synthetic diamonds that are more flawless than natural diamonds, "metal rubber," medical diagnostic devices 1,000 times faster and 100,000 times more accurate than today's state-of-the-art products, scores of new cancer drugs that promise to kill cancer cells long before they ever pose a threat to human health, and computer chips that are not manufactured in the traditional sense but rather are grown atom-by-atom.

As a result, virtually every industry from energy and semiconductors to pharmaceuticals and manufacturing will face a seismic shift equal, if not greater, than that experienced by the Swiss watch industry.

None of the companies listed in this chapter are yet publicly traded, but investors need to be aware of them for two reasons. First, many of them are either likely to go public or be acquired by public companies. Therefore, investors who track their progress can either get in on the initial public offering (IPO) or, if the company is acquired, take an equity stake in the acquiring company. Secondly, investors need to understand how these companies and their technologies could change the "rules of the game" for entire industrial sectors. For instance, investing in traditional oil, gas, and coal energy companies may look like a safe, attractive option today, but if nanotechnology enables radically more efficient fuel cells and solar cells, the future of energy production could look entirely differently. The same is true for the semiconductor industry. The world is undoubtedly going to continue to need more powerful computer chips for the foreseeable future, but there is no reason why those chips have to be manufactured in the traditional sense (requiring sophisticated multibillion dollar fabrication facilities) when they can be grown cheaply and effectively in a standard lab.

As has been stated in other chapters, these companies and their technologies must be viewed realistically. Not all of them will succeed. Some have not yet perfected their technology, and as such, potential customers may be reluctant to adopt their technologies until they can be demonstrated to be 100 percent effective. (After all, if your health is at stake or you are counting on a new chip to operate vital aspects of your business, you are unlikely to accept a drug or chip that is anything less than perfect.) Still other companies must prove they can build their products at commercially scalable levels.

Lastly, investors must understand that many of the large corporations—especially those who are threatened by the new technology—are unlikely to go down without a fight. They may develop their own competing technologies, attempt to acquire (and stifle) the new technologies, or use their marketing and lobbying power to slow its advance. That, however, is the nature of business.

The one thing we know is that change is inevitable. What follows is a list of those nanotechnology companies most likely to bring us the next great waves of changes.

The Disruptors

COMPANY	Angstrom Medica
INVESTORS	Individual investors
ADDRESS	150-A New Boston Street Woburn, MA 01801
PHONE	781-933-6121
CEO	Paul Mraz
WEB	*www.angstrommedica.com*

DESCRIPTION   Angstrom Medica is a biomaterials company engaged in the development and commercialization of its patented nanocrystalline phosphate technology—NanOss.

WHY IT IS DISRUPTIVE   Human bone is made of a calcium and phosphate composite called hydroxyapatite. By manipulating calcium and phosphate at the molecular level, Angstrom Medica has created a material that is identical in structure and composition to natural bone. This suggests that NanOss, because it can facilitate the rate at which these molecules are absorbed by bone cells, can speed the repair and healing of broken and damaged bones. Furthermore, because the nanocrystals Angstrom Medica manufacturers are so small (less than 100 nanometers), there are fewer places for stress to build up, which makes the material just as strong as regular bone.

WHAT TO WATCH FOR   Angstrom Medica is competing with Orthovita, and although it received almost $4 million in funding in 2004, it will need to initiate clinical trials soon in order to gain FDA approval and get a viable product on the market within a reasonable time frame.

COMPANY	4Wave
INVESTORS	Milestone Equity Partners, LLC
ADDRESS	22977 Eaglewood Court, Suite 120 Sterling, VA 20166
PHONE	703-787-9033
CEO	Sami Antrazi (President)
WEB	*www.4wave.com*

DESCRIPTION   4Wave has developed expertise in manufacturing materials that are created one atomic layer at a time. The technology, which is called biased target ion beam deposition, is used to create optical films for multifilter chips.

WHY IT IS DISRUPTIVE   4Wave's chips are capable of combining four wavelengths, each capable of transmitting 2.5 gigabytes of data per second over a fiber-optic cable, and then deciphering the light at the end of the transmission to get the data out. What makes the technology so exciting is that compared with today's optical filters, 4Wave's technology will be 250 times smaller, more reliable, and up to 80 percent less expensive to manufacture.

WHAT TO WATCH FOR   4Wave has received $2 million from the National Institute of Standards and Technology to create its chips. If the company can land new customers (IBM, Hitachi, and Seagate are reportedly testing the technology), it will serve as a validation of its technology. The company's stated goal is to receive $45 million in revenue by 2008. It is an ambitious goal because it faces competition from other nanotechnology start-ups such as NanoOpto and NeoPhotonics.

COMPANY	Apollo Diamond, Inc.
INVESTORS	Private investors
ADDRESS	60 State Street, Suite 700 Boston, MA 02110
PHONE	508-881-4060
CEO	Bryant Linares
WEB	*www.apollodiamond.com*

DESCRIPTION   Apollo Diamond grows 100 percent diamond crystals—through a process called chemical vapor disposition—that will match or exceed the purity of the world's finest naturally mined diamonds.

WHY IT IS DISRUPTIVE   Apollo is currently growing diamonds for less than $100 per carat. What makes these diamonds superior to naturally mined diamonds is that they are absolutely flawless. (Mined diamonds usually have some minor imperfections.) The technology has the potential to end DeBeers 115-year monopoly of the diamond industry.

WHAT TO WATCH FOR   As promising as the retail diamond market is, Apollo is more interested in producing diamond wafers that can be used in next-generation computer chips. Today's state-of-the-art computer chips are running hotter and hotter. Eventually, the chips will run so hot that they will liquefy silicon. Diamondoid materials have superior thermal properties and could be the next material of choice for the massive semiconductor industry. If Apollo can crack this market, it will be a very bullish sign.

The Disruptors

COMPANY	Argonide Corporation
INVESTORS	Private
ADDRESS	291 Power Court Sanford, FL 32771
PHONE	407-322-2500
CEO	Fred Tepper
WEB	*www.argonide.com*

DESCRIPTION   Argonide manufacturers and commercializes water purification systems employing nanomaterials.

WHY IT IS DISRUPTIVE   The company's NanoCeram technology removes viruses, bacteria, and endotoxins not on the basis of size, but rather electrical charge. This allows its nanofilters to block 99.9999 percent of all viruses at water flow rates several hundred times greater than conventional filters. The worldwide market for water management products is an estimated $400 billion.

WHAT TO WATCH FOR   Argonide has received a number of government grants to help develop its products. It now needs to demonstrate success in the commercial marketplace. It faces competition from smaller companies such as KX Industries, TriSep, Seldon Laboratories, and NanoMagnetics, as well as larger ones such as GE Water, Donaldson Company, Pentair, and Calgon Carbon.

The Disruptors

COMPANY	BioForce Nanoscience, Inc.
INVESTORS	Gulf Stream Capital, Societe Generale Asset Management, Iowa First Capital Fund
ADDRESS	2901 South Loop Drive, Suite 3400 Ames, IA 50010
PHONE	515-296-6550
CEO	Dr. Laurence Russ
WEB	*www.bioforcenano.com*

DESCRIPTION   BioForce Nanoscience develops ultraminiaturized nanoarray technologies for solid phase, high-throughput biomolecular analysis.

WHY IT IS DISRUPTIVE   The company's nanoarrays not only possess the ability to test a larger number of samples than a convention mircoarray but it can also reportedly do it 10 to 100 times faster and less expensively. Furthermore, because its nanoarrays are read with atomic force microscopes instead of biomarkers (like dyes or quantum dots), they don't alter the sample in any way. The ability to test molecules, materials, and cells will allow BioForce's product to be used for material development, biomedical diagnostics, and drug discovery.

WHAT TO WATCH FOR   The company began delivering products in 2005 and has estimated that it will be profitable by 2007. If leading research institutions and major pharmaceutical companies begin purchasing its equipment, it will be a very good sign. Although it is more likely to be acquired by a larger competitor, if the company does go public, investors should consider an investment.

COMPANY	Cambrios Technologies
INVESTORS	Arch Ventures, Alloy Ventures, Oxford Bioscience Partners, Lux Capital, Avalon Ventures, Harris & Harris, In-Q-Tel
ADDRESS	2450 Bayshore Parkway Mountain View, CA 94043
CEO	Michael R. Knapp, Ph.D.
WEB	*www.cambrios.com*

DESCRIPTION   Founded on the basis of Dr. Angela Belcher's (a 2004 MacArthur Fellowship award winner) research, the company is developing proteins that can bind to a wide variety of different electronic, optical, and magnetic materials as well as self-assemble other materials in small protein structures.

WHY IT IS DISRUPTIVE   Belcher and her colleagues at Cambrios are essentially attempting to mimic how Mother Nature constructs materials. If they are successful, semiconductors, nanowires, and scores of other small electrical and optical components won't be manufactured in the traditional sense; they will be grown. In addition to having the potential to be atomically precise, such devices would have the advantage of being manufactured through simplified, energy-efficient processes that don't require large, multibillion dollar fabrication facilities.

WHAT TO WATCH FOR   Cambrios is still a research company, and it needs to find partners to help commercialize its technology. If it can successfully develop real products (i.e., thin films, nanowires, etc.), this could be one of the more disruptive nanotechnology companies because it will have introduced an entirely new manufacturing paradigm. Such a process could dominate how the next generation of nanoelectronics and other nanoscale devices are built. The company is working on over thirty other short-term projects, including how to self-assemble magnetic storage materials, solar cells, and flexible batteries.

COMPANY	Dendritic Nanotechnologies
INVESTORS	Starpharma, Dow Chemical
ADDRESS	2625 Denison Drive Mount Pleasant, MI 48858
PHONE	989-774-4199
CEO	Robert Berry, PhD
WEB	*www.dnanotech.com*

DESCRIPTION Dendritic Nanotechnologies develops dendritic polymer delivery technology for use in diagnostic, therapeutic, and drug delivery solutions.

WHY IT IS DISRUPTIVE Dendrimers are a new type of polymer that can be precisely synthesized with specific properties—everything from size to conductivity. These unique properties, which also include the ability to hold molecules, means they have potential applications as new drug delivery devices. The Defense Department, through Dendritic's relationship with the U.S. Army's Institute for Soldier Nanotechnologies, is also exploring their use in topical emulsions to detoxify chemical and biological agents. In May 2005, the company announced it had produced a new line of dendrimers called the Priostar. What makes the Priostar so disruptive is that it can be manufactured in two to three days (as opposed to thirty days) for a fraction of the cost. These new economics will open up a wide variety of new uses for dendrimers.

WHAT TO WATCH FOR Dendritic is already selling over 200 different types of dendrimers through its Web site and its relationship with Sigma-Aldrich. But the real proof of dendrimers' value will be if they demonstrate success as a drug delivery device (the approval of Starpharma's VivaGel will be the first test) and if they can be produced reliably and in large enough quantities to be used as a platform to battle chemical and biological agents. If they do these, look for the company to enter into some partnerships or licensing arrangements with some large companies. In early 2005, Dow Chemical agreed to exchange its portfolio of dendrimers patents for an equity stake in the company. This leaves Dendritic Nanotechnologies (and Starpharma) as the world leaders in this field. NanoBio (see page 216), however, is also pursuing dendrimers for the treatment of cancer.

COMPANY	Fluidigm
INVESTORS	Versant Ventures, Lehman Brothers Healthcare Fund, Piper Jaffray Ventures, In-Q-Tel
ADDRESS	7100 Shoreline Court South San Francisco, CA 94080
PHONE	650-266-6000
CEO	Gajus Worthington
WEB	*www.fluidigm.com*

DESCRIPTION   Fluidigm manufacturers integrated fluidic circuits (IFCs) that can manipulate biological samples at the molecular level through a vast network of valves, channels, vials, and reaction chambers on a single chip.

WHY IT IS DISRUPTIVE   Fluidigm's IFCs could do for the life sciences/biotechnology sector what integrated circuits did for computing and the manipulation of electrons— which is rapidly and inexpensively manipulate biological samples. Fluidigm's chips are 100 times faster than other "labs-on-a-chip," less expensive, and more accurate. They are also among the first to place the equivalent of a full-blown experimentation laboratory on a single chip.

WHAT TO WATCH FOR   Fluidigm already counts among its customers Merck, Eli Lilly, and GlaxoSmithKline, but if it moves into the stem cell area or if its chips become even more proficient at detecting the presence of proteins that predict cardiovascular disease or Alzheimer's, it will be a very bullish signal. Fluidigm also has a relation- ship with the Central Intelligence Agency to develop its technology for genetic fin- gerprinting and the early detection of biological and chemical threats. If the company goes public, investors should give it serious consideration. (The company's technology appears superior to that of Nanostream, which in spite of its name, is not a nanotech- nology company.)

COMPANY	The Gemesis Corporation
INVESTORS	Private
ADDRESS	7040 Professional Parkway East Sarasota, FL 34240
PHONE	941-907-9889
CEO	Carter Clarke
WEB	*www.gemesis.com*

DESCRIPTION   Gemesis grows gem quality diamond crystals by applying the proper pressure and temperature to create an environment conducive for common carbon to rearrange its atoms into rough diamond crystals (which then must be cut and polished into its final form). The company specializes in manufacturing yellow and orange diamonds.

WHY IT IS DISRUPTIVE   The yellow and orange diamonds that Gemesis manufactures cost approximately $100 per carat and currently retail for between $10,000 and $15,000. The price difference and the fact that they cannot be distinguished by the human eye make the diamonds disruptive to the diamond industry.

WHAT TO WATCH FOR   Unlike Apollo Diamond's manufactured diamonds—which are flawless—Gemesis' diamonds can be distinguished from naturally mined diamonds by testing equipment. As a result, the Federal Trade Commission has ruled that the company could not call them real diamonds. Gemesis has chosen to label them "cultured diamonds." Gemesis' technology is very good and so are its products. The big question that needs to be determined is how will consumers respond to the product. Specifically, will people still want naturally mined diamonds, or will they be willing to buy Gemesis' manufactured diamonds?

The Disruptors

COMPANY	iMEDD, Inc.
INVESTORS	Vernon International, Angel investors
ADDRESS	1381 Kinnear Road, Suite 111 Columbus, OH 43212
PHONE	614-340-6001
CEO	Carl F. Grove
WEB	*www.imeddinc.com*

DESCRIPTION iMEDD is an early stage biomedical company developing micropar-
ticles and nanomembranes capable of releasing drug dosages into the body at a slow,
steady, and sustained rate. Its NanoGate interferon product is targeting hepatitis C.

WHY IT IS DISRUPTIVE iMEDD's nanomembrane technology can create highly uniform pore sizes (in the 10 to 100 nanometer range) that can control drug diffusion
at the molecular level. For example, one possible platform it could create is a "bio-
capsule" that could protect islet cells (which are necessary for creating an artificial
pancreas in a diabetic patient). Because the pore size can be regulated, it could be
precisely designed to allow in the glucose molecules and the other nutrients necessary to produce the insulin, but large enough to keep out harmful antibodies. If the
technology is approved by the FDA, it could free up diabetic patients from having to
take insulin shots as often.

WHAT TO WATCH FOR iMedd's products are still a long way from FDA approval. It
also faces competition from competing technologies, but it bears watching, especially
if its NanoGate product is approved.

COMPANY	Integrated Nano-Technologies
INVESTORS	Angel investors
ADDRESS	999 Lehigh Station Road Henrietta, NY 14467
PHONE	585-334-0170
CEO	D. Michael Connolly
WEB	*www.integratednano.com*

DESCRIPTION   Integrated Nano-Technologies fuses molecular biology, chemistry, and microelectronics in an attempt to create a variety of self-assembled nanoscale circuits. The company has already produced a field-portable DNA identification systems that can determine if chemical or biological compounds (such as anthrax) are present in the environment.

WHY IT IS DISRUPTIVE   The company is working to improve its technology to test for up to sixty-four different pathogens or diseases on a single chip. Because the process is inexpensive and biocompatible, it can potentially be used in everything from biosecurity, diagnostics, RFIDs, medical devices, and food safety.

WHAT TO WATCH FOR   Short term, watch if the company develops a prototype field system and then achieves large-scale manufacturing by early 2007. Longer-term, if Integrated Nano-Technologies can figure out a way to use DNA as a method for guiding the self-assembly of circuits, it could potentially challenge Cambrios for developing an entirely new paradigm for fabricating next-generation circuits.

The Disruptors

COMPANY	Kereos, Inc.
INVESTORS	Angel investors
ADDRESS	4041 Forest Park Avenue St. Louis, MO 63108
PHONE	314-633-1879
CEO	Robert A. Beardsley
WEB	*www.kereos.com*

DESCRIPTION   Kereos, Inc., develops targeted therapeutics and molecular imaging agents that detect and attack cancer and cardiovascular disease faster and with more specificity than other agents and drugs.

WHY IT IS DISRUPTIVE   Kereos K1-0001—an MRI agent for tumor detection—can detect tumors as small as 1mm or about ten times smaller than today's best agents. This means that preventative treatments can begin much sooner and give patients a better shot at full recovery. Kereos is currently partnering with Bristol Myers Squibb to develop these agents for detecting atherosclerotic plague—the leading cause of heart attacks. The company is also developing targeted chemotherapeutics to address a number of different cancers that are more potent and less toxic than those currently on the market.

WHAT TO WATCH FOR   Keroes' relationships with Bristol Myers Squibb and Dow Chemical are suggestive of the potential of its technology but the clinical trials are paramount. Keoreos is currently pursuing trials on drugs to address America's top two killers—cancer and heart disease. If either trial goes well, look for a number of major pharmaceutical companies to try to partner with the company. Depending on the positive outcome of these trials, if Kereos goes public, investors should consider an investment.

COMPANY	Konarka Technologies
INVESTORS	Draper Fisher Jurvetson, Zero Stage Capital, Chevron-Texaco, Eastman Chemical, Mitsui and Onpoint Technologies
ADDRESS	100 Foot of John Street Booth Mill South, Third Floor, Suite 12 Lowell, MA 01852
PHONE	978-569-1400
CEO	Howard Berke
WEB	*www.konarkatech.com*

DESCRIPTION   Konarka utilizes nanomaterials and conductive polymers to manufacture light-activated power plastics that are inexpensive, lightweight, flexible, and versatile.

WHY IT IS DISRUPTIVE   Today, less than 1 percent of the world's energy is derived from solar power. However, when the economic, environmental, and geopolitical aspects of traditional energy sources—coal, oil, natural gas, and nuclear power—are taken into consideration, solar power becomes an attractive choice, especially if its price can come down. What Konarka is doing is developing plastic rolls that can be embedded with titanium dioxide nanoparticles that can efficiently convert both natural and indoor light into electricity. The company first hopes to develop flexible solar coatings for laptop computers and mobile phones. Longer term, it is working on inexpensive, lightweight rolls of solar cells that could cover a home or business, thus freeing occupants from having to be connected to the existing power grid. If successful, not only could the United States's energy production change radically, but also millions of people in the developing world would have access to cheap, clean, and sustainable energy.

WHAT TO WATCH FOR   BP Solar, Nanosys, and Nanosolar are all working on related technologies. There is no guarantee Konarka will develop the best or most practical product, but its partnerships are all very promising. If cell phone manufacturers begin incorporating Konarka's technology into next-generation devices or it enters into a partnership with a major manufacturer to produce solar "shingles," it will be a positive sign. The most positive indicator, however, will be the company's ability to decrease the cost of energy produced per watt. Investors should definitely keep this company on their radar screen for a future IPO.

COMPANY	KX Industries
INVESTORS	McGowan Capital
ADDRESS	269 South Lambert Road Orange, CT 06477-3502
PHONE	800-462-8745
CEO	Evan Koslow
WEB	*www.kxindustries.com*

DESCRIPTION   KX Industries develops and sells a variety of carbon filters employing nanofibers.

WHY IT IS DISRUPTIVE   Like Argonide, KX Industries is seeking to employ nanotechnology—in this case nanofibers—to filter out a large percentage of viruses, bacteria, and microbes in drinking water. The technology has the potential to provide safe drinking water to the developing world.

WHAT TO WATCH FOR   The distribution of its technology into the developing world is the key to its success. The fact that KX has supplied filters to Proctor & Gamble for its Pur and Brita brands is promising and suggests it has a leg up on competitors like Argonide. However, Seldon Laboratories, another competitor, appears to have received a greater share of government contracts to develop its filters. (It is working with NASA to develop a water recirculation purification system.)

COMPANY	NanoBio Corporation
INVESTORS	Private
ADDRESS	2311 Green Rd, Suite A Ann Arbor, MI 48105
PHONE	734-302-4000
CEO	Michael J. Nestor
WEB	*www.nanobio.com*

DESCRIPTION NanoBio Corporation is a biopharmaceutical company specializing in the development of antimicrobial nanoemulsion technology to treat a variety of diseases, including herpes, cold sores, and nail fungus. In late 2003, it created a separate company, NanoCure, to develop dendrimers for the treatment of cancer.

WHY IT IS DISRUPTIVE In 2004, NanoBio conducted FDA-approved Phase I clinical trials for NanoHPX, a topical treatment for herpes. If the clinical trials prove successful (the results were not yet available at the time of publication), it could provide a cure for cold sores—an outbreak that affects an estimated 50 million people every year. The company is also working on treatments for SARS and AIDS.

NanoCure's technology is even more promising because it holds the potential to diagnose, image, and deliver a cancer-killing agent to an individual cancer cell with a single dendrimer device. Such a platform would revolutionize cancer treatment and could portend the day when cancer cells are eradicated at a very early stage. The company has also received a $3.2 million grant from the Defense Department to develop nanoemulsions capable of neutralizing and detoxifying biological and chemical agents.

WHAT TO WATCH FOR If the company receives approval from the FDA to proceed to Phase IIA or Phase II clinical trials for NanoHPX, it will be a bullish sign. Also, NanoCure is a direct competitor to Starpharma's dendrimer technology. Whichever company is the first to receive FDA approval for its product will have a first-mover advantage—although partnerships with major pharmaceutical companies will be the most important factor in determining long-term success.

COMPANY	Nanochip
INVESTORS	JK&B Capital, New Enterprise Associates, Microsoft, AKN Technology
ADDRESS	48041 Fremont Boulevard Fremont, CA 94538
PHONE	510-770-2501
CEO	Gordon Knight
WEB	*www.nanochip.com*

DESCRIPTION   Nanochip is developing a new class of storage products (which employ microelectrical mechanical systems and nanoscale atomic probes) that are smaller, faster, use less power, and achieve higher storage densities than today's state-of-the-art flash memory devices.

WHY IT IS DISRUPTIVE   Today's current flash memory can store about 50 gigabits. Nanochip claims it will be able to store 100 gigabits with the potential to go as high as 50,000 gigabits per square inch. If it can successfully manufacture a product with even a fraction of the later figure, it could capture a large portion of today's $12 billion market for flash memory.

WHAT TO WATCH FOR   The technology has not yet been proven. It is currently in beta trials with customers. If the customers are pleased with the product and demand more, look for Nanochip to license its technology to a large manufacturer who can help with production and distribution.

COMPANY:	Nanoconduction Inc.
INVESTORS:	Jerusalem Venture Partners and the Woodside Fund.
ADDRESS:	1275 Reamwood Avenue Sunnyvale, CA 94089
PHONE:	408-702-1000
CEO:	Bala Padmakumar
WEB:	*www.nanoconduction.com*

DESCRIPTION   Nanoconduction is designing and developing a process—and the equipment—to use carbon nanotubes to remove heat from today's start-of-the art semiconductor circuits.

WHY IT IS DISRUPTIVE   As semiconductor circuits continue to shrink in size, heat management is becoming an increasingly serious issue. Nanoconduction has stated that its approach can deliver a three-fold improvement in conducting heat away from computer chips. The company is currently working with NASA to develop a prototype.

WHAT TO WATCH FOR   Nanoconduction has already licensed some NASA-developed material (a carbon nanotube array composite) but it must refine its process so that it can produce enough units to warrant the time and attention of major semiconductor manufacturers. To this end, the announcement of a partnership—such as the one Zyvex (see page x) has with Intel—would be noteworthy.

CONCLUSION   Nanoconduction faces a good deal of competition from companies like NanoCoolers and Zyvex but it does bear watching because, unlike its competitors, its process uses multi-walled carbon nanotubes instead of single wall nanotubes. This is important because the former are easier to synthesize (and thus more economical) than the latter. However, until Nanoconduction actually announces a partnership with a major semiconductor company, it just another company with a promising but unproven technology.

COMPANY	NanoCoolers, Inc.
INVESTORS	Draper Fisher Jurvetson, Austin Ventures, and others
ADDRESS	5307 Industrial Oaks Boulevard, Suite 100 Austin, TX 78735
PHONE	512-583-8000
CEO	Jim Moore
WEB	*www.nanocoolers.com*

DESCRIPTION   NanoCoolers is developing advanced cooling technology that employs both liquid metals and thin film thermoelectrics to cool computers and other devices.

WHY IT IS DISRUPTIVE   As today's state-of-the-art computers get smaller and more powerful, they are also getting hotter. By replacing water with liquid metal—which is sixty-five times as thermally conductive as water, nontoxic, and environmentally safe—NanoCoolers hopes to employ its technology in next-generation computer processors. The company's technology is already being tested by Hewlett-Packard and IBM.

NanoCoolers' thin film thermoelectrics are equally disruptive in that they can lead to solid-state refrigeration devices that support high cooling densities, are more reliable (they have no moving parts), and have smaller form factors. These advantages could lead to smaller, and more effective and efficient, refrigerators. It could also lead to the development of smaller transportable coolers, instant icemakers, and water chillers that are attached directly to faucets.

WHAT TO WATCH FOR   In the short term, NanoCoolers is going after the computer market. If IBM, Intel, or Hewlett-Packard begins using its technology, it will be a bullish sign. Longer term, the possibilities in the refrigeration market are just as great—especially if the company's technology can lead to the redesign of more space-efficient refrigerators.

COMPANY	NanoDynamics
INVESTORS	Angel investors
ADDRESS	901 Fuhrmann Boulevard Buffalo, NY 14203
PHONE	716-853-4900
CEO	Keith Blakely
WEB	*www.nanodynamics.com*

DESCRIPTION   NanoDynamics develops and commercializes nanomaterials for a wide variety of applications from everything from fuel cells to golf balls.

WHY IT IS DISRUPTIVE   NanoDynamics Revolution 50, a solid oxide fuel cell (SOFC), uses nanopowders to improve the efficiency of the device. The product was slated to hit the commercial marketplace in late 2005, and because SOFC can operate on hydrocarbons (such as propane), it is more practical than a typical fuel cell (which uses only hydrogen) and can be used to power everything from military field gear and battery recharger systems to outdoor billboards and refrigeration systems. In late 2004, the company also introduced its NDMX golf ball, which it claims is not as susceptible to hooking and slicing because the ball's molecular structure reduces how the ball spins. If approved by the Professional Golfer's Association, it could claim a large share of the golf ball market.

WHAT TO WATCH FOR   The commercial success of the Revolution 50 will go a long way toward determining whether NanoDynamics is successful. But because its potential product pipeline for its nanomaterials is so large and varied, it is not entirely dependent on its fuel cell. The problem is that in the other nanomaterials areas it faces a good deal of competition from companies such as Nanophase and Oxonica.

COMPANY	NanoHorizons, Inc.
INVESTORS	Angel investors
ADDRESS	Technology Center, Suite 208
	200 Innovation Boulevard
	State College, PA 16803
PHONE	814-861-9909
CEO	Robert F. Burlinson
WEB	*www.nanohorizons.com*

DESCRIPTION   NanoHorizons is developing and employing nanotechnology to produce mass spectrometry equipment, flexible electronics, chemical sensors, and nanoparticles.

WHY IT IS DISRUPTIVE   Its mass spectrometry technology holds the potential to help pharmaceutical companies analyze drug candidates in minutes—as opposed to hours. Its flexible electronics technology is being explored by companies like pSivida (see pages 190–91) to assist in the areas of diagnostics and tissue engineering. Lastly, the company has garnered some favorable press by touting its nanoparticles as having the ability to produce odor-free and microbial-free socks and shoes.

WHAT TO WATCH FOR   NanoHorizons holds the license to a good deal of intellectual property from Penn State University. It remains to be seen, however, whether the company can actually translate any of the promise of its technology into commercial products. To date, no commercial buyers have stepped forward for either its mass spectrometry equipment, flexible electronic technology, or its nanoparticles. Furthermore, in each area the company faces competition from both larger companies as well as better financed private start-ups.

COMPANY	Nanoplex
INVESTORS	Angel investors
ADDRESS	665 Clyde Avenue Mountain View, CA 94043
PHONE	650-230-1589
CEO	Michael Natan
WEB	*www.nanoplextech.com*

DESCRIPTION   Nanoplex develops metal nanoparticles with unique properties that allow them to be used for ultrasensitive detection in life science research and in tagging to provide authentication and tracking of consumer goods.

WHY IT IS DISRUPTIVE   Nanoplex's nanoparticles really act like tiny "nanobarcodes" and can be incorporated into ink, fabric, clothing, paper, jewelry, and even explosives to ensure brand security. Inventory shrinkage (e.g., consumer theft) and counterfeiting costs the industry billions of dollars every year. Nanoplex's nanobarcodes could eliminate this problem.

WHAT TO WATCH FOR   The life-science applications for nanobarcodes are also very significant because genomics and proteomics are fueling the demand for rapid, simultaneous detection and measurement of multiple biomolecules in a variety of different samples. They may also be very helpful in detecting cancer cells in small samples of blood or tissue. If Nanoplex can get the price of its nanobracodes down to an appropriate price point, it could be poised for some very nice growth.

The Disruptors

COMPANY	NanoOpto Corp.
INVESTORS	Harris & Harris Group, Draper Fisher Jurvetson, Morganthaler Ventures, and others
ADDRESS	1600 Cottontail Lane Somerset, NJ 08873-5117
PHONE	732-627-0808
CEO	Barry Weinbaum
WEB	*www.nanoopto.com*

DESCRIPTION   NanoOpto has designed proprietary nanofabrication technology and nanoimprint lithography to produce a broad range of unique optical components, including switches and polarizing filters for the telecommunications industry and sub-components for the digital imaging and disk drive markets.

WHY IT IS DISRUPTIVE   If successful, NanoOpto's optical components will enable higher-quality, lower-cost optical systems and provide for better integration among a variety of consumer and commercial products, and in spite of the telecommunications industry's troubles over the past few years, the demand for ever smaller, cheaper, and more functional electro-optic devices will increase for the foreseeable future.

WHAT TO WATCH FOR   To the extent that NanoOpto can actually produce products that are reliable and cost competitive, it should do well. In the short term, look for it to enter the CD and DVD markets. The fact that the company has already signed manufacturing and licensing deals with a few companies is a promising sign. NanoOpto faces a lot of competition, and the market for its products is a demanding one.

COMPANY	Nanosolar
INVESTORS	Benchmark Capital, U.S. Venture Partners
ADDRESS	2440 Embarcadero Way Palo Alto, CA 94303-3313
PHONE	650-565-8891
CEO	Martin Roscheisen
WEB	*www.nanosolar.com*

DESCRIPTION   Nanosolar has developed and is producing nanowires, nanoparticles, and self-assembling nanomaterials that will allow for the creation of flexible, low-cost photovoltaic cells.

WHY IT IS DISRUPTIVE   Traditional solar cells are manufactured out of silicon, which is costly, bulky, and inflexible. Nanosolar's technology (as well as Konarka's and Nanosys') offers the possibility that its solar cells will be 1,000 times thinner and can be manufactured 100 times faster. In 2005, the company received a $10 million grant from DARPA to develop its technology, and it holds some valuable intellectual property from Sandia National Laboratories. The technology may even lend itself to being "painted" on the sides of automobiles and buses. If this is achieved, it could potentially allow a wide variety of products to generate their own energy.

WHAT TO WATCH FOR   The market for solar cells is growing at an annual rate of 40 percent per annum. Nanotechnology will likely continue and may even add to this impressive growth. Longer term, if its solar cells can produce electricity for less than the cost of today's grid, the entire energy paradigm will have shifted, and the nearly $1 trillion energy market will be up for grabs. The company will face competition not only from Konarka and Nanosys, but also from Cypress, Miasole, General Electric, and Sharp (the largest manufacturer of silicon solar cells).

COMPANY	NanoSonic, Inc.
INVESTORS	Private
ADDRESS	1485 South Main Street Blacksburg, VA 24060
PHONE	540-953-1785
CEO	Rick Claus
WEB	*www.nanosonic.com*

DESCRIPTION   NanoSonic is developing a revolutionary molecular self-assembling process to design new materials. One of the company's new materials is something called Metal Rubber, which, as its name suggests, has the high electrical conductivity of a metal along with the flexibility of rubber.

WHY IT IS DISRUPTIVE   NanoSonic has already received funding from a variety of government agencies, and it has reportedly signed some type of agreement with Lockheed Martin. Metal Rubber has a number of innovative uses that may challenge today's producers of steel, aluminum, composites, or plastics. Possible uses of the material include artificial muscles, "smart" clothing (i.e., clothing with electronics embedded in it), and even shape-shifting airplane wings. In short, the technology could change the nature of the age old question of form versus function.

WHAT TO WATCH FOR   NanoSonic currently doesn't possess the ability to manufacture Metal Rubber in industrial-level quantities. If this changes or if it partners with a major manufacturer, the company's revenues could increase significantly. Also watch for news coming out of the partnership with Lockheed Martin. Other bullish signs will include if NanoSonic begins licensing its technology to other major manufacturers or if its technology is employed by major medical device manufacturers.

COMPANY	NanoSpectra Biosciences
INVESTORS	Angel investors
ADDRESS	8285 El Rio Street, Suite 130 Houston, TX 77054
PHONE	713-842-2720
CEO	J. Donald Payne
WEB	*www.nanospectra.com*

DESCRIPTION   NanoSpectra is developing a unique therapy for the destruction of various cancers using its patented nanoshell particles. In 2004, it received a $2 million grant from the Advanced Technology Program to explore its nanoshell technology.

WHY IT IS DISRUPTIVE   By developing precisely tailored nanoparticles, NanoSpectra can potentially make drugs that are absorbed only by tumorous cells. Once absorbed, the nanoparticles—because they can be tuned to either absorb or scatter light at desired wavelengths—can be heated with infrared light to the point where they are killed. Such a technology has the potential to become a valuable tool in the treatment of cancer (the second leading cause of death in the United States). It also suggests that chemotherapy in the future may become obsolete. More recently, research has suggested that nanoparticles can be tailored to enhance chemical sensing by as much as 10 billion times. By determining how light scatters off molecules, scientists can better determine the chemical makeup of molecules and materials. This could significantly improve the scientific and medical community's ability to understand how cells, genes, and proteins operate at the biological level.

WHAT TO WATCH FOR   NanoSpectra has not yet begun the FDA approval process, meaning that its technology is still years away from the marketplace. The novelty of its technology alone, however, suggests that the company bears watching.

COMPANY	Nanosphere, Inc.
INVESTORS	Lurie Investments, NextGen Partners, TakaraBio
ADDRESS	4088 Commercial Avenue Northbrook, IL 60062
PHONE	847-400-9000
CEO	William P. Moffitt
WEB	*www.nanosphere-inc.com*

DESCRIPTION   Nanosphere is a nanotechnology-based life sciences company that is developing a universal molecular testing system.

WHY IT IS DISRUPTIVE   Nanosphere's technology has the potential to be 1,000 times more sensitive than traditional molecular testing. It is also easier to use, has a quicker turnaround time, and is capable of multiplexing (testing for more than one molecular marker at a time). The company's nanoparticles can attach to DNA, RNA, or proteins and quickly detect their presence. The technology is superior to today's polymerase chain reaction (PCR) technology that amplifies the molecules in DNA until a sufficient quality is present to allow detection. Nanosphere's technology holds the potential of making molecular testing more practical by making smaller and more flexible testing devices. This, in turn, could help reduce health care costs by enabling doctors to detect a patients' predisposition to certain medical conditions at a much earlier stage. It could also improve patient care by optimizing patient drug response based on genetic variation and enhance public safety by detecting various biological and chemical agents, such as anthrax and plague, in miniscule amounts at an early stage. For instance, the technology could be used at airport screeners or to test the nation's water supplies.

WHAT TO WATCH FOR   Nanosphere's technology has already been tested in a handful of hospitals around the country and has received favorable reports. If its assay technology receives FDA approval, look for it to be extremely competitive with the medical diagnostic technology of Nanogen, Cepheid, and Roche.

COMPANY	The NanoSteel company	
INVESTORS	Military Commercial Technologies	
ADDRESS	485 North Keller Road, Suite 100	
	Maitland, FL 32751	
PHONE	407-838-1427	
CEO	Joseph Buffa	
WEB	*www.nanosteelco.com*	

DESCRIPTION   The NanoSteel Company develops nanostructured steel.

WHY IT IS DISRUPTIVE   NanoSteel's nanostructured steel and its nanocomposites coatings create materials that, due to the low density of defects, are capable of bearing external stress and physical payloads at far greater levels than conventional materials. The materials also have greater wear, corrosion, and impact resistance. The nanomaterials could lead to everything from knives that never dull and car frames that can absorb the impact of a major collision to stronger bridges and the creation of new materials that will allow for new architectural applications.

WHAT TO WATCH FOR   The military has funded and tested NanoSteel's material. If the company can produce its materials reliably and at a reasonable price, it should be able to gain the attention of scores of manufacturers and find niches all along the material supply chain. The company, however, faces competition from companies such as MMFX Microcomposite Steel.

COMPANY	NanoString Technologies
INVESTORS	Draper Fisher Jurvetson, OVP Venture Partners
ADDRESS	201 Elliott Avenue West, Suite 300 Seattle, WA 98119
PHONE	206-378-6266
CEO	H. Perry Fell
WEB	*www.nanostring.com*

DESCRIPTION   NanoString is developing a patent-pending nanotechnology-based platform for high-speed, highly multiplexed single molecule identification and digital quantification.

WHY IT IS DISRUPTIVE   NanoString's technology claims to be able to apply a unique bar code to each individual targeted molecule in a biological sample and then scan that information and create an inventory of every molecule. It claims to be faster, cheaper, and more sensitive than microarray technology and, if successful, could revolutionize gene analysis and lead to an era of predictive, preventative, and personalized medicine.

WHAT TO WATCH FOR   NanoString is a very early stage company. Until it actually demonstrates some success—such as helping to diagnosis cancer at a very early stage—or it attracts the attention of leading microarray companies (such as Agilent and Affymetrix), investors should simply keep their eye on the company.

The Disruptors

COMPANY	Nanosys, Inc.
INVESTORS	Venrock, ARCH Venture Partners, Harris & Harris, Lux Capital, SAIC Venture Capital Group, and others
ADDRESS	2625 Hanover Street Palo Alto, CA 94304
PHONE	650-331-2100
CEO	Calvin Chow
WEB	*www.nanosysinc.com*

DESCRIPTION   Nanosys, Inc. is an industry-leading nanotechnology company that is developing a wide variety of products based on a technology platform incorporating high performance inorganic nanostructures. It currently holds over 250 patents and has already established partnerships with some of the world's leading companies, including Intel, Dupont, Matsushita, and H.B. Fuller in the fields of nonvolatile memory, flexible electronics, low-cost nanocomposite solar cells, fuel cell technology, and nanocoatings.

WHY IT IS DISRUPTIVE   Nanosys' nanocomposite solar cells (which it has entered into an agreement with Matsushita to begin producing in 2007) will compete with Konarka and Nanosolar's technology and could usher in an era of clean, affordable alternative energy. It is also partnering with Sharp Corporation to develop nanotechnology-related fuel cells that could be used in portable electronic devices and thus reduce the need for batteries. The company's thin film electronics technology could be used in everything from flat panel displays and antenna arrays for wireless communications to low-cost radio frequency identification tags and "smart" clothing. Nanosys' work with Intel in the area of nonvolatile memory could lead to cheaper but more capable MP3 players, digital cameras, and mobile phones. Lastly, it is also constructing nanowires that could lead to faster, cheap, and highly effective biosensors.

WHAT TO WATCH FOR   The company could attempt to go public in 2006. Investors should keep an eye on whether it can move beyond its current reliance on government contracts. The production of its solar cells with Matsushita is also a key benchmark. If Nanosys stays on track and the company begins selling its solar cells, it will be a very positive sign. Nanosys' intellectual property alone is valuable; however, it can only become a blockbuster company if it actually begins producing real products.

The Disruptors

COMPANY	Nano-Tex LLC
INVESTORS	Owned by Burlington Industries
ADDRESS	5770 Shellmound Street Emeryville, CA 94608
PHONE	510-420-3772
CEO	Donn Tice
WEB	*www.nano-tex.com*

DESCRIPTION   Nano-Tex is a specialty chemical company that focuses on designing molecules to attach to clothing and other fabrics during the milling process.

WHY IT IS DISRUPTIVE   The nanoparticles Nano-Tex employs are used to create clothing that is wrinkle-free, water-repellent, and stain-resistant. In 2004, over 20 million garments incorporated Nano-Tex's nanotechnology. As the number of production partners continues to grow (it currently has established relations with Eddie Bauer, Lee Jean, DKNY, Tommy Hilfiger, Gap, Old Navy, and fifty other manufacturers worldwide), the demand for cleaning products, like detergent, and cleaning services, such as dry cleaners, will slowly but steadily decline.

WHAT TO WATCH FOR   In addition to continuing to line up more clothing manufacturers in the United States and abroad, look for Nano-Tex to expand its presence into a variety of other textile uses, including furniture fabrics and mattresses. The company is also reportedly developing Nano-Touch, which will give synthetic materials a texture like cotton. Longer term, look for foreign competitors to develop technologies to compete with Nano-Tex.

COMPANY	Nantero, Inc.
INVESTORS	Draper Fisher Jurvetson, Harris & Harris, Stata Venture Partners, and others
ADDRESS	25-D Olympia Avenue Woburn, MA 01801
PHONE	203-656-0833
CEO	Greg Schmergel
WEB	*www.nantero.com*

DESCRIPTION   Nantero is working with carbon nanotubes to develop nonvolatile random access memory (NRAM)—a portable memory chip with low power consumption, ultrahigh storage density, and a speed high enough to compete with—and potentially replace—all other forms of memory.

WHY IT IS DISRUPTIVE   Nantero has successfully demonstrated that it can properly align over 10 billion carbon nanotubes on a single wafer. It also claims that its technology has been incorporated into a standard semiconductor production line, suggesting that major chip manufacturers will not have to alter their production processes to employ its technology. If Nantero is successful, its technology could replace all existing forms of memory and lead to more capable laptops, cell phones, and other electronic devices, as well as "instant-on" computers.

WHAT TO WATCH FOR   Nantero has stated that it will have a product by late 2005. If the company's tests with LSI Logic (who is sampling its technology) prove successful, look for other major semiconductor manufacturers such as IBM, Motorola, and Infineon to enter into licensing agreements with Nantero. Be aware, however, that these companies and other companies are also working on competing technologies. Recently, Nantero also announced a partnership with BAE System—a major aerospace manufacturer—which signaled Nantero's move into competition with Eikos (see page 86) who is also working with carbon nanotubes to combat the effects of radiation.

The Disruptors

	COMPANY	NeoPhotonics
	INVESTORS	Oak Investment Partners, Institutional Venture Partners, Draper Fisher Jurvetson, Harris & Harris, Rockport Capital Partners, and others
	ADDRESS	2911 Zanker Road San Jose, CA 95131
	PHONE	408-232-9200
	CEO	Tim Jenks
	WEB	*www.neophotonics.com*

DESCRIPTION NeoPhotonics develops and manufacturers advanced planar optical devices by integrating active and passive optical materials using the company's proprietary nanomaterials-based processes. The company's intellectual property gives it a strong foundation to build next-generation optical systems.

WHY IT IS DISRUPTIVE In spite of the stagnation in the telecommunications industry over the past few years, companies are still trying to employ various technologies to increase the data throughput of fiber—especially in metropolitan areas. NeoPhotonics, using a process called Laser Reactive Deposition, can now deposit particles less than 20 nanometers in size on almost any material. By improving a variety of optical components, NeoPhotonics believes it will be able to integrate more functions into a smaller space at a lower cost on these components.

WHAT TO WATCH FOR In addition to competing with NanoOpto, NeoPhotonics is also up against JDS Uniphase Corp., Alcatel, and Nortel Networks. If the company can successfully create cost-competitive products that enable fiber-to-the-premise (FTTP) deployment and make broadband fiber access possible for homes and small businesses, NeoPhotonics could experience large growth. Although the company is more likely to be acquired by a larger competitor if this happens, should it instead go public, it may offer an attractive investment for the more risk tolerant investor.

COMPANY	Ntera Ltd.
INVESTORS	Cross Atlantic Capital Partners, Doughty Hanson Technology Ventures
ADDRESS	58 Spruce Street Stillorgan Industrial Park Dublin, Ireland
PHONE	353.1.213.7564
CEO	Nick How
WEB	*www.ntera.com*

DESCRIPTION   Ntera has developed an electrochromic display technology using nanostructured film electrodes.

WHY IT IS DISRUPTIVE   Ntera's nanostructured films, which it markets as NanoChromics, allow up to 1,000 times as many electrochromic molecules to be placed on the surface area of a display, giving the screen a much more vivid color. The high speed of the molecules' electron transfer gives the technology a very high switching speed. The technology is expected to compete with Liquid Crystal Displays in terms of picture quality and lower energy consumption.

WHAT TO WATCH FOR   Ntera is a small company and the fact that so many other major corporations are also making products to compete with LCDs—in the form of carbon nanotube-based flat panel displays—makes it doubtful the company will succeed. If, however, the company can go after niche markets—such as the screens for clocks and instruments displays—it may find a road to profitability.

The Disruptors

COMPANY	QuesTek Innovation
INVESTORS	Angel investors
ADDRESS	1820 Ridge Avenue Evanston, IL 60201
PHONE	847-328-5800
CEO	Charles Kuehmann
WEB	*www.questek.com*

DESCRIPTION   QuesTek is a material solutions company that employs proprietary software to design and develop new materials, including nanomaterials.

WHY IT IS DISRUPTIVE   QuesTek can rapidly and inexpensively develop high-performance steels and other alloys in about 60 percent of the time and at about 75 percent of the cost of traditional manufacturing methods. This suggests that the company may find a wide variety of high value-added applications for its materials. The company is already working with NASA and the Defense Department on a number of aerospace-related projects. Nearer term, the company may be able to find markets in the automotive racing, cutlery, and sporting goods fields by making those products stronger and lighter.

WHAT TO WATCH FOR   The U.S. Air Force is currently evaluating QuesTek's Ferrium53 alloy for its landing gears, and the U.S. Navy is exploring another of its alloys for carrier-based aircraft components. If either alloy receives certification, it will serve as a significant validation of QuesTek's technology.

COMPANY	Seldon Laboratories, LLC
INVESTORS	Private and government grants
ADDRESS	7 Everett Lane, Door 18 Windsor, VT 05089
PHONE	802-674-2444
CEO	Alan Cummings
WEB	*www.seldontech.com*

DESCRIPTION   Seldon Laboratories is focusing on employing carbon nanotubes into a variety of products. At the present time, its main application is the creation of carbon nanotube-based filters.

WHY IT IS DISRUPTIVE   Seldon has received a $2 million grant from NASA to develop a space-based water purification system. It has also received a $600,000 grant from the U.S. Air Force to produce a portable water purification system. It is disruptive technology because it does not rely on chemicals (like chlorine) or ultraviolet light to kill microorganisms, making filters easier and cheaper to use. If the company is successful, its technology could free up soldiers from having to carry water with them on dangerous missions. (They could simply filter any water that is available.) The technology also has applications for campers, hikers, and, of course, for the hundreds of millions of people in the developing world who live in areas where potable water is scarce.

WHAT TO WATCH FOR   Seldon is in direct competition with KX Industries and Argonide. Whichever company can develop the most effective and cost-competitive technology, as well as enter into partnerships with companies who can help distribute its technology, will be the most likely to win in the commercial marketplace.

The Disruptors

COMPANY	Sequoia Pacific Research Company, LLC
INVESTORS	Private
ADDRESS	1192 E. Draper Parkway, Suite 522 Draper, UT 84020
PHONE	801-303-2558
CEO	Richard L. Maile
WEB	*www.sequoiaprc.com*

DESCRIPTION   Sequoia Pacific is a chemical research and development holding company specializing in the development of nanoengineered organic material to address a variety of environmental issues.

WHY IT IS DISRUPTIVE   Sequoia has already developed one product—SoilSET—that is capable of binding with soil to protect it from erosion and thus stimulate agricultural growth. Because the nanoscale particles bind with the soil and help retain water, the nanomaterials can help in everything from reclaiming land lost to wild fires to reducing runoff and germinating seeds.

WHAT TO WATCH FOR   Because Sequoia is a research and development company, it will need partnerships to succeed. Moreover, while it has successfully sold its products to the Bureau of Indian Affairs to help reclaim land lost to a fire in New Mexico, the much larger market for its product will be in the area of stimulating agricultural growth. If it can increase the productive value of agricultural land, it will likely find a large market for its product.

COMPANY	SoluBest
INVESTORS	Alplex BV, Israeli Nanoparticle Consortium
ADDRESS	18 Einstein St. Ness Ziona 74140 Israel
PHONE	972.8.940.3023
CEO	Erwin Stern
WEB	*www.solubest.com*

DESCRIPTION   SoluBest develops and re-engineers drug molecules to create new drug formulations with favorable physical and pharmaceutical profiles.

WHY IT IS DISRUPTIVE   SoluBest's technology has the potential to improve the solubility and bioavailability of both new and existing drugs. The latter is very significant because half of the world's top sixty drugs will come off patent by 2008. SoluBest's technology can potentially extend the patent protection by reformulating the drug.

WHAT TO WATCH FOR   SoluBest is competing with Flamel, SkyePharma, and Elan. If the company is successful (and it is reportedly working with three "big pharma" companies), it could hurt the other companies' prospects. Also, if the company's technology is successful, look for a big pharma company to acquire SoluBest. Should it go public, investors are encouraged to consider an investment.

The Disruptors

COMPANY	ZettaCore, Inc.
INVESTORS	Kleiner Perkins Caufield, Draper Fisher Jurvetson, Oxford Biosciences, Access Ventures, and others
ADDRESS	369 Inverness Parkway, Suite 350 Englewood, CO 80112
PHONE	303-300-2900
CEO	Subodh Toprani
WEB	*www.zettacore.com*

DESCRIPTION   ZettaCore is developing molecular memory technology and products for current and next-generation semiconductors.

WHY IT IS DISRUPTIVE   By manipulating the electronic properties of molecules and taking advantage of their self-assembling properties, ZettaCore promises to be able to enable very high-density memory devices. Its technology can add or remove electrons from single molecules, and then those states can be read as individual bits of information. Moreover, because the company claims to be able to get molecules to self-assemble on a substrate (like silicon), it promises to create not only very dense memory, but it will also be inexpensive to manufacture. Such a technology could revolutionize the whole electronics industry.

WHAT TO WATCH FOR   ZettaCore believes its technology will soon be able to be used to improve existing semiconductor memory. If this technology receives validation from a major semiconductor manufacturer, it will be a very significant milestone because it will suggest that its molecules are stable at higher temperatures and do not degrade after trillions of read-write cycles. NVE Corporation (see pages 187–88) and Nantero (see page 232), as well as the likes of Hewlett-Packard and IBM, are all competing in this same field.

COMPANY	Zia Laser
INVESTORS	Harris & Harris, Venrock Associates, RWI Group, and Prism Venture Partners
ADDRESS	801 University Blvd SE, Suite 105 Albuquerque, NM 87106
PHONE	505-243-3070
CEO	Kenneth Westrick
WEB	*www.zialaser.com*

DESCRIPTION   Zia Laser is a leading innovator and manufacturer of quantum dot semiconductor lasers.

WHY IT IS DISRUPTIVE   Today, microprocessors have hit a wall in terms of speed. Zia Laser's technology allows computer chips to process some of its signals optically instead of electronically. This could lead to significantly faster, more efficient, and less expensive microprocessors. If successful, the technology could create a new generation of microprocessors.

WHAT TO WATCH FOR   Zia Lasers has reportedly received some funding and is working with a "leading" semiconductor manufacturer. If the relationship matures and the company becomes public, it will be a bullish signal. Investors are definitely encouraged to keep Zia Laser on their radar screens.

The Disruptors

COMPANY	Zyvex Corporation
INVESTORS	Private and government grants
ADDRESS	1321 North Plano Road Richardson, TX 75081
PHONE	972-235-7881
CEO	James Von Ehr
WEB	*www.zyvex.com*

DESCRIPTION   Zyvex is developing a variety of nanoscale characterization and manipulation tools and structures, as well as nanomaterials to facilitate molecular manufacturing.

WHY IT IS DISRUPTIVE   In 2002, the company received a $25 million grant from NIST to develop a prototype assembler employing microelectromechanical systems (MEMS) for the purpose of developing a nanoelectromechanical system (NEMS). If it is successful, Zyvex will be well along the road to developing equipment that could aid in the manufacturing of devices at the molecular level. The practical development of such a tool could quite literally turn the manufacturing industry (and virtually every other industry) on its head. Rather than scaling materials down to make ever smaller components (like today's semiconductor chips), the new paradigm would be to build things from the atom up. This capability would decrease size, reduce waste, increase functionality, and lead to a variety of applications not yet conceived.

On a more practical level, Zyvex is also certifying carbon nanotube suppliers and applying them to a host of practical uses. For instance, it is using carbon nanotubes to improve the mechanical strength of a variety of composites used in the sporting goods industry; Intel Corporation is experimenting with them to enhance the thermal properties of its computer chips; and NASA is exploring their potential to create high-strength, low-weight nanocomposites.

WHAT TO WATCH FOR   Zyvex has been very successful in gaining government grants to support its research and development. To its credit, however, it has also begun to line up commercial vendors and even reported $2 million for the first quarter of 2005. If Intel (or another major manufacturer) becomes a major purchaser of its carbon nano-tubes, it will be a positive sign for the short to midterm. Longer term, its work in the area of NEMS and molecular manipulation promises the most exciting opportunities.

Summary

From materials that can help regrow human bone to "metal rubbers" to medical diagnostic technologies hundreds to thousands of times faster than today's state-of-the-art technology, the nanotechnology companies in this chapter are at the forefront of a revolution that will have significant implications for investors—whether or not they are paying attention to the field.

As you move forward, always be mindful of these companies—especially if they go public—because they do not just represent exciting investment opportunities, they also portend an age of "creative destruction" unlike anything witnessed to date.

For instance, Apollo Diamond could disrupt not just the entire diamond industry; it could also replace a significant portion of the silicon industry and Cambrios—by growing electronic components instead of manufacturing them in the traditional sense—or Zia Laser could usher in paradigm shifts of historic proportions in the semiconductor industry.

The risks are great, but so are the potential rewards. For those investors willing to stay close to shore and keep their eye on the horizon, the possibility that they will be able to catch one of the next great waves—a wave that will catch many others almost completely unaware—is a very real possibility.

"For a successful technology, reality must take precedence over public relations, for Nature cannot be fooled."

—Richard Feynman

Chapter 8

Forewarned Is Forearmed: Common Dangers and Risky Companies

Fact: Some of the companies listed in this book will not be in existence in five years. A few are even likely to be out of business by the time this book hits the bookshelves. The reality is that, even in spite of doing due diligence, there are things that are beyond the capability to anticipate, predict, or know. This truism makes the stock market what it is. For every person that thinks a stock is a good investment and is willing to buy it at a given price, someone feels the opposite and is willing to sell it.

With that dose of cold water, here are the ten greatest dangers associated with investing in nanotechnology stocks. The chapter will conclude with a handful of companies that I believe either demonstrate or are at risk of falling prey to some of these dangers.

Danger #1: Too Many Competitors

Capitalism thrives on competition and is, arguably, at the root of much of the economic prosperity the developed world has experienced over the past century. Competition is also a dual-edged sword in that the successful companies prosper at the expense of the vanquished. The reality is that there are a number of companies who are applying almost identical technologies to the same market. For instance, numerous companies are exploring how to utilize carbon nanotubes in flat panel displays. Others are seeking to apply quantum dots to manufacture LEDs, and still others are manufacturing a variety of nanomaterials with nearly identical properties. For instance, 3M, Othovita, and Angstrom Medica are all developing nanomaterials to facilitate the regrowth of human bone. The market is such that it can likely support only one of the company's technologies. The others will be forced to find new applications or be relegated to the proverbial ash heap.

Danger #2: Nanotechnology Offers More Solutions Than There Are Problems and Will Fragment the Market

Put in more common terms, nanotechnology usually offers more than one way to skin most cats. Throughout this book, numerous companies have been profiled that are applying slightly different approaches to the same problem. For instance, a number of companies are working on promising alternative energy platforms. Nanosys, Konarka, and NanoSolar are developing flexible solar cells, while NanoDynamics and others are utilizing nanotechnology to improve the effectiveness of fuel cells. One or more of these companies might meet with significant commercial marketplace success, or they may all find select niches.

An alternative scenario, however, is that nanotechnology will be employed to make coal cleaner and nuclear power safer and thus negate the need for the wide-scale use of either solar cell or fuel cell technology.

The bottom line is that all or some of the technologies may succeed, and to the degree that they do, the market share for any given technology may be smaller than originally anticipated. This isn't necessarily bad. Nice profits can still be had in certain niches, but a company's stock price can suffer if the market size decreases too much.

Danger #3: Poor Management

One of nanotechnology's allures is that the unique physical, chemical, and biological properties that are inherent at the nanoscale open up markets in a variety of different fields. For instance, many companies speak of applying nanomaterials or nanosensors to the energy, pharmaceutical, manufacturing, and agricultural sectors. Often, the company is correct in noting that its technology does have potential applications in a number of markets; however, the problem stems from the fact that market realities often dictate the company focuses on one field first and develop a specific product that meets a real need for that market. Too many companies though—especially small start-ups—want to do it all and they often lack the strategic focus necessary to succeed.

Another problem, which has been evidenced too much over the past few years, is corporate malfeasance. Because nanotechnology is in its infancy and because it is still generally not understood by the public makes it more likely that less-than-scrupulous individuals will attempt to profit off of people's ignorance. Such executives and their companies may survive for a time, but as the last stock bubble demonstrated, their only lasting legacy will be the trail of investor misery they leave in their wake.

Danger #4: Lack of Financial Resources

The old saying "It takes money to make money" is true. Few companies are profitable from the beginning, and most take at least a few years to become so. Most early stage investors recognize this fact and are, in exchange for getting in at an attractive price, willing to demonstrate some patience. But most investors have limited patience. Many of the companies listed in this book are the recipients of venture capital. Venture capitalists are typically willing to give a company anywhere between three and seven years to prove they can either become a viable acquisition target or succeed on the public market. If companies cannot demonstrate progress toward these established goals, funding is going to become increasingly difficult to obtain. The farther away a company is from manufacturing a reliable product in scalable amounts, the more difficult it'll be to get funding. For publicly traded companies, the issuing of additional shares should generally be taken as a warning sign unless the company can clearly articulate why diluting shareholder value is in the long-term interest of the investors.

Another common source of funding for nanotechnology start-ups is government grants. This is an attractive source of early money because it often funds necessary research and development, and the grants don't have to be repaid—meaning it doesn't dilute stockholder equity. Investors, however, need to be cautious of companies whose only outside funding is from government grants.

Danger #5: Better Is Not Always Good Enough

A great many nanotechnologies can and will make products better. The problem is that these improvements often come with a large price increase. Sometimes people are willing to pay a high premium for even a marginal improvement in a product. For instance, the military will likely pay top dollar to ensure that the latest and greatest nanocomposites are installed on its next-generation jet fighter; and health care patients often will pay more for a product that only incrementally improves their health for the simple reason they don't want to skimp when it comes to their well-being.

Many other products, however, don't have the same advantage. For instance, a number of companies are manufacturing self-cleaning windows and self-cleaning ceramics. These products are nice, and consumers will likely purchase them if they don't come with too steep a price. The same can be true for nanotechnology-enabled LED light bulbs. These devices might well last ten to twenty times longer than an average lightbulb, and they could even be a wonderful long-term investment. But consumers aren't always logical. To understand, look at today's battery market. Certain batteries last four times longer than average batteries but are three times as expensive. From an economic perspective, they make a wise purchase. Unfortunately, for the manufacturers, consumers are still reluctant to buy them because many are unable to grasp the long-term savings.

Danger #6: Change Is Hard

It is human nature to resist change. People do it, and even large corporations do it. Many of the companies profiled in this book will require people and organizations to change their habits or behavior. For instance, a number of companies are working on drug delivery platforms that could allow any number of drugs to be released into a person's body according to a prescribed time frame. Such a device would free up people who must take multiple drugs from remembering which, when, and how many drug tablets they must take. The devices, however, must be implanted in the person. It may well be a better and more effective way to administer drugs, but if people are uncomfortable with the idea of giving up control over their drugs to an inanimate object, it may not be accepted by consumers.

Large corporations are equally resistant to change. Often there are legitimate business reasons for this. Many of their customers rely on their products for their health or the operation of their business, and anything less than perfection is not acceptable. Because of this, corporations are unlikely to change—let alone tinker with—processes and products that have proven successful. Nantero, NVE, and many others are working on technologies that could well revolutionize the computer industry, but until these companies can demonstrate they are 100 percent reliable, the large manufacturers are unlikely to accept their technology.

It must also be remembered that many of these companies have sizeable investments in their current infrastructure and may not want to change their processes or retrain their workers on how to operate new equipment.

Danger #7: Beware the Lawyers

For a field as complex as nanotechnology, the absence so far of any serious lawsuit battles may seem surprising. The reason for this is that until recently few companies were making any money in the field of nanotechnology. Lawsuits, especially lawsuits over intellectual property, often make sense from an economic perspective only if one side feels there is money to be made. Now that profits, in some cases big profits, are on the near-term horizon, investors can expect this to change.

In early 1991, a researcher at NEC discovered the first carbon nanotube. It was not until 2004 that the company announced its intention to enforce the patent. The reason is simple. Carbon nanotubes are now moving into mainstream production, and NEC stands to profit handsomely if either some of its competitors stop their work altogether or, alternatively, are required to negotiate a licensing agreement with NEC to manufacture and sell carbon nanotubes.

Until the issue is settled, it will hang like a rain cloud over the industry's head. Then depending on how it is resolved, it will transform the industry.

The same scenario could play out over a variety of nanotechnologies. Up until late 2004, the U.S. Trade and Patent office didn't even have a separate code for nanotechnology patents. The implication is that many of the thousands of nanotechnology-related patents issued over the past decade may have been very broadly issued, or because the patent examiners didn't fully understand the technology, similar or overlapping patents may have been issued. Under either scenario, the ensuing fights could well be played out in the courts over the years ahead and at great expense to the litigants.

Investors need to understand that smaller companies, especially those with limited resources, are at a significant disadvantage in a fight with a large corporation who has deep pockets. The bottom line is that such a fight isn't always won by those in the right; they are won by those with the most financial might and the better lawyers.

Danger #8: Beware the Regulators

Nanotechnology has a real need to be properly regulated. The field is creating nanomaterials that are new to society. How these materials interact with the natural environment and in the human body are legitimate concerns. For instance, nanoparticles, due to their small size, may react with cells in unknown ways or cross the blood-brain barrier and have serious unintended consequences.

These scenarios are neither a given, nor can they be dismissed out of hand. The problem for investors is that until the nanomaterials can be deemed safe a significant number of nanotechnology investments remain at risk. In 2004, one study indicated that carbon nanotubes had damaged the brains of certain fish. This study was later found to have some methodological flaws, but until further studies can prove beyond a reasonable doubt that it isn't the case, every company involved in the manufacture or use of carbon nanotubes carries a risk.

Other nanomaterials, nanoparticles, and nanoscale devices will require approval from the Food and Drug Administration. For drugs, FDA approval is a long and timely process. Therefore, many promising companies, such as Starpharma, Elan, and Flamel, who currently have nanoparticle or dendrimers drugs in clinical trials, remain risky as well.

Investors will also need to heed whether certain nanoscale devices are to be regulated as a drug or as a medical device. Dendrimers are a perfect example. These synthetic branchlike devices can be tailor-made to hold any number of molecules or antibodies. If the FDA rules each unique dendrimer is a drug, that will slow their development. If, however, they are ruled to be a device, their application is likely to be more quickly and more broadly felt in the pharmaceutical sector.

Lastly, investors will need to beware of how else a product might be regulated. Chapter 1 spoke about how the diamond industry has successfully gotten the Federal Trade Commission to require one of the companies manufacturing diamonds to label their diamonds as synthetic. Other nanotechnology-related products could be regulated in a similar fashion.

Danger #9: Commoditization

Carbon nanotubes, nanoparticles, quantum dots, and a variety of new nanocomposites and nanomaterials all have unique and valuable properties. All of these properties can be exploited to make existing products better, as well as make entirely new products and devices. These enhanced properties and new devices may carry a premier price in the beginning, but as companies scale up their production levels and as more competitors get into the field, the price of these nanoscale materials and devices could drop. Thus, companies that were growing and profitable one year may find their profits evaporate over a relatively short period of time.

Danger #10: "Gray Goo," "Prey," and the Real Problems

In the last few years, Prince Charles and others have either suggested or called for an outright ban on nanotechnology. Among the more popular reasons suggested for such a ban have been a series of articles and books discussing "self-assembling nanobots" that could potentially destroy all matter in its path and reduce it to a pile of disassembled atoms that resemble "gray goo." This, in a nutshell, is the plot of Michael Crichton's 2003 bestselling novel, *Prey*. Without going into detail, the science behind his scenario is faulty, and the book is fiction—not fact. Still, the public could easily latch onto this and other doomsday scenarios, and the resultant negative public perception of nanotechnology could easily dim enthusiasm for nanotechnology companies or kill a great deal of promising nanotechnology research and development.

To its credit, the nanotechnology community, along with the U.S. government, is doing a great deal to handle and address in a responsible and forthright manner the legitimate environmental and health questions that nanotechnology poses.

The more serious issues that could slow nanotechnology relate to the host of very real societal and ethical issues to which the technology will give rise. For instance, there is little doubt that nanotechnology will disrupt certain large industries and that workers will be displaced. It is a serious issue and unions and those adversely affected by the change could attempt to slow certain nanotechnology advances. Another serious concern is how nanotechnology will affect wealth creation—and destruction. Using the earlier example of the new synthetic diamonds, if diamonds continue to be cheaply produced in massive quantities, what happens to the appraised value of all of today's valuable diamonds? Lastly, nanotechnology advances in medicine are certain to transform life expectancy projections. How does society care and provide for an older population? What happens to Social Security and Medicare? Until such questions are satisfactorily answered, a certain amount of risk will remain to every nanotech investment.

Red Flags!

With all of these sobering factors in mind, there are a handful of companies that highlight some of the aforementioned dangers and of which investors will want to be very cautious.

ALTI	COMPANY	Altair Nanotechnologies, Inc.
	SYMBOL	ALTI
	TRADING MARKET	NASDAQ
	ADDRESS	204 Edison Way Reno, NV 89502
	PHONE	775-858-2500
	CEO	Dr. Alan Gotcher
	WEB	*www.altairnano.com*

DESCRIPTION   Altair describes itself as a manufacturer of unique nanocrystalline materials. The company believes its proprietary products have a variety of applications in solar cells, drug delivery, fuel cells, and advanced batteries.

REASONS TO BE BULLISH

► Altair has demonstrated some success landing modest contracts from the U.S. government.

► In 2005, it signed a partnership with Advanced Battery Technologies to use its nanomaterials in its lithium-ion battery technology.

REASONS TO BE BEARISH

► The company is generating minimal revenue and is expecting to lose close to $6 million in 2005.

► Its leading product, RenaZorb, which is designed to control phosphate levels in kidney dialysis patients, has not found any customers, and its other products— TiNano Spheres (used for drug delivery) and Titanium Dioxide pigment—both face serious competition from more experienced competitors.

► Altair's past announcements of new products have failed to deliver on promises. For instance, in 2003, the company announced the development of its NanoCheck Algae Preventer and predicted the product would find a niche in the pool/spa arena. No sales have yet to develop.

continued

Altair Nanotechnologies, Inc. continued

WHAT TO WATCH FOR   Be aware of press releases touting new products that the company claims will capture large or very lucrative markets. Investors should also not be persuaded by news of the company being awarded modest government grants. Only announcements that demonstrate it is actually receiving real revenue from commercial customers should be accorded any significance.

CONCLUSION   Outlook is very bearish. In 2004, Altair's stock price fluctuated between $1 and $4.40. Even at the low end, with almost 50 million shares outstanding, the company had a market capitalization of $50 million. This is extraordinarily high for a company with only $640,000 in revenues. Investors should also treat this stock with great caution because Altair's management team seems to lack strategic focus. (In the past few years, it has claimed to be pursuing products in the fuel cell, drug delivery, and material sciences segments.) It also faces very tough competition from more established and better funded competitors such as Oxonica, Nanophase, and 3M.

NNPP	COMPANY	Applied Nanotech, Inc./Nano-Proprietary
	SYMBOL	NNPP
	TRADING MARKET	Over-the-counter Bulletin Board
	ADDRESS	3006 Longhorn Boulevard, Suite 107 Usa, TX 78758
	PHONE	512-339-5020
	CEO	Marc Eller
	WEB	*www.nano-proprietary.com*

DESCRIPTION   Applied Nanotech is a wholly-owned subsidiary of Nano-Proprietary, Inc. The company specializes in the development and licensing of carbon nanotubes and Carbon Field Emission (CFE) products.

REASONS TO BE BULLISH

► Applied Nanotech claims patents for several nano-emissive display technologies.

► The company recently restructured its business to focus exclusively on licensing its intellectual property, and it has reportedly entered into an agreement to license its technology for the manufacturing of hydrogen sensors (although the company is unnamed).

► Its carbon nanotubes have been demonstrated to work in a prototype flat panel display.

REASONS TO BE BEARISH

► Nano-Proprietary had only $68,000 in revenue for the first quarter of 2005 and reported a loss of $1.5 million.

► The company will likely need to generate additional financing in late 2005, and it will possibly do so by issuing additional shares of the company's stock.

► It faces stiff competition, especially in the race to incorporate their carbon nanotubes in flat panel displays, from other carbon nanotube producers including NEC and Carbon Nanotechnologies.

continued

Applied Nanotech, Inc./Nano-Proprietary continued

WHAT TO WATCH FOR   Nano-Proprietary has already sued a few larger companies, including Canon, over claims that it is infringing on its patents. If the company prevails, it could generate a significant stream of revenue in the form of licensing agreements. On the other hand, in order to defend these claims, the company will likely have to issue additional shares to finance the litigation. If it does, look for the stock to drop even further.

CONCLUSION   Outlook is very bearish. In 2004, the company's stock traded between $1.25 and $3.25—giving it a market capitalization of between $121 million and $315 million. For a company with minimal revenues, the price would appear to be unsustainable. The downside risk is substantially greater than the possible upside potential.

Risky Companies

NNLX		
	COMPANY	Nanologix
	SYMBOL	NNLX
	TRADING MARKET	Over-the-counter Bulletin Board
	ADDRESS	87 Stambaugh Avenue, Suite 2 Sharon, PA 16146
	PHONE	724-346-1302
	CEO	Mitchell S. Felder
	WEB	*www.infectech.com*

DESCRIPTION   Formerly known as InfecTech, Nanologix claims to be a technology innovator of biogas products and services to assist in worldwide energy interdependence and reduce pollution by employing renewable, environmentally-friendly energy sources. More recently, it moved into the research and development of diagnostic test kits to help identify infectious human diseases.

REASONS TO BE BULLISH
► None at the present time.

REASONS TO BE BEARISH
► No known revenues and no customers.
► In both the areas of energy production and disease detection, it faces formidable competition from better-established and better-financed companies.

WHAT TO WATCH FOR   The company claims to be working on a technology to produce hydrogen cheaply but investors should treat this with great caution. Investors should also be wary of other press releases that either tout large markets or claims that it is working with "unnamed" big companies.

CONCLUSION   Outlook is very bearish. Until Nanologix has an actual product and then some customers, investors should stay away from the stock.

Risky Companies

NANO	COMPANY	NanoMetrics, Inc.
	SYMBOL	NANO
	TRADING MARKET	NASDAQ
	ADDRESS	1550 Buckeye Drive Milpitas, CA 95035
	PHONE	408-435-9600
	CEO	John Heaton
	WEB	*www.nanometrics.com*

DESCRIPTION   In spite of its name and its ticker symbol (NANO), NanoMetrics is not involved in nanotechnology as defined in this book. The company manufactures and markets process control metrology systems for the semiconductor and flat panel display industries. In January 2005, it announced that it would be merging with August Technology, a company in a related field, and taking the new name of August NanoMetrics.

CONCLUSION   Because the company is not involved with nanotechnology, no analysis has been conducted. But investors should not be confused by the company's ticker symbol.

NPCT.BB	COMPANY	NanoPierce Technologies, Inc.
	SYMBOL	NPCT.BB
	TRADING MARKET	Bulletin Board
	ADDRESS	370 17th Street, Suite 3640 Denver, CO 80202
	PHONE	303-592-1010
	CEO	Paul Metzinger
	WEB	*www.nanopierce.com*

DESCRIPTION   Little extensive information is available regarding NanoPierce. As best determined, the company is attempting to produce a device it calls the Nano-Pierce Connection System (NCS) which it claims will provide for the superior production of circuit boards, radio frequency identification tags (RFIDs), sockets, and other electronic systems.

REASONS TO BE BULLISH
► None.

REASONS TO BE BEARISH
► The company had no known sales or revenues in 2004 or the first part of 2005.
► It has consistently failed to deliver on past promises. For example, in 2002, the company announced that it expected sales of its new electricity conductor to surge, but to date, no sales have yet been reported.
► The NanoPierce management team would appear to lack expertise in the field of nanotechnology.
► The fact that the company has changed names from Sunlight Systems and Mendell-Denver Corp. to NanoPierce raises concerns that it is simply playing to investor interest in nanotechnology.
► Its announcement in late 2004 that it was going to produce, market, and sell "a biotech yeast beta glucan product" appears to be a significant departure from its previous market of nanoelectronics and raises additional concerns.

continued

NanoPierce Technologies, Inc. continued

WHAT TO WATCH FOR   Be alert for news releases that tout new products that the company claims will lead to significant revenues. Be especially wary of such announcements if they correspond to periods of heightened public interest in nanotechnology.

CONCLUSION   Outlook is very bearish. In 2004, the company's stock traded between 9 cents and 75 cents per share—giving it a market capitalization of between $16 million and $134 million. For a company that has consistently failed to produce any revenues and whose management and market strategy is suspect, it appears to represent only a downside risk.

Risky Companies

USGA.OB	COMPANY	US Global Nanospace
	SYMBOL	USGA.OB
	TRADING MARKET	Bulletin Board
	ADDRESS	2533 North Carson Street, Suite 5107 Carson, City 89706
	PHONE	775-841-3246
	CEO	John Robinson
	WEB	*www.usgn.com*

DESCRIPTION   US Global Nanospace, formerly known as US Global Aerospace and before that as Caring Products International, bills itself as a nanomaterials company specializing in the development of antiballistic material sciences. It also reportedly specializes in the development of "optimized polymer and organic materials and nanofibers . . . for superior filtration system for air, water and cigarettes, biological and chemical decontaminants, and blast mitigation and fire protection."

REASONS TO BE BULLISH
► No known reasons at the present time.

REASONS TO BE BEARISH
► For 2004, the company reported only $40,000 in revenues and a loss of nearly $17 million.
► A close review of the company's 10Q filing revealed some disturbing quotes, such as "we do not have the infrastructure or financial resources to manufacture and distribute our products" and "our products have not been proven consistent for commercial use."
► The company also has a history of reporting large sales that have not materialized. (For example, the company claimed to have sold 200 Guardian Cockpit Security doors to El Al Israel Airlines but apparently did not.)

WHAT TO WATCH FOR   If one of the company's press releases sounds too good to be true, based on the company's previous track record, it probably is. Investors are encouraged to look only at the company's revenues and its profits (if it has any).

continued

Risky Companies

US Global Nanospace continued

CONCLUSION Outlook is very bearish. The company has a history of issuing impressive-sounding press releases—which have driven the stock price up—but have constantly failed to deliver on the news reported in their releases. In the past year, the company's stock has traded between 12 cents and $3.05. The latter price gave it a market capitalization of $245 million. As a result, when the price dropped, a number of uninformed investors lost a significant amount of money in the process.

Summary

In the early 1980s, one of the most popular television shows was *Hill Street Blues.* In every show, the lead police sergeant would gather the beat cops together for a briefing before they hit the streets. He would always conclude his remarks with the same line: "And, hey, let's be careful out there." It was good advice then, and it is good advice for investors getting into nanotechnology. As with any investment, caution and common sense are the watchwords.

"When companies like Merrill Lynch start having a nanotechnology index, I think that is getting into the hype cycle for which a lot of people got into trouble during the Internet bubble."

—Vinod Khosla, Venture Capitalist

Chapter 9

Tracking Nanotechnology and Creating Your Own Nanotech Mutual Fund

On April 1, 2004, Merrill Lynch released the first ever nanotechnology index. It was offered as a tool to help investors evaluate the nanotechnology "space." Within one week, the company had to sheepishly remove a handful of companies from its list because they had nothing to do with nanotechnology. Suddenly, the irony of the release date—April's Fool Day—should have been apparent to all.

It wasn't. In June 2004, Piper Jaffray, not wishing to be upstaged by Merrill Lynch, released its own report on nanotechnology and listed some of the very companies that Merrill Lynch had already retracted from its index. Worse, it lumped in a number of other very dubious nanotechnology companies in its index. Piper Jaffray offered no qualitative analysis on any of these companies listed, and their clients could easily have been left with the impression that one company was as much a "nanotechnology pure play" as another. (To the company's credit, the report did advise clients not to construe the list as a recommendation, but readers could be forgiven if they didn't catch that advice after the glowing introduction about the possibilities of nanotechnology.)

At the same time, a third firm, Punk Ziegel Company, created yet another index. Both Merrill Lynch's and Punk Ziegel's indexes are listed as follows. Like Piper Jaffray, neither Merrill Lynch nor Punk Ziegel suggest its indexes should be used as guidance for selecting nanotechnology stocks. Rather, they are intended to serve as a proxy for the broader nanotechnology "market."

MERRILL LYNCH NANOTECH INDEX

Accelrys Software, Inc.
Altair Nanotechnologies, Inc.
Amcol International Corp.
BioSante Pharmaceuticals, Inc.
Cabot Corp.
CombiMatrix Group
FEI Co.
Flamel Technologies
Harris & Harris Group, Inc.
Headwaters, Inc.
Immunicon Corp.
JMAR Technologies, Inc.

Kopin Corp.
MTS Systems Corp.
Nanogen, Inc.
Nanophase Technologies
Novavax, Inc.
NVE Corp.
SkyePharma PLC
Symyx Technologies, Inc.
Tegal Corp.
Ultratech, Inc.
Veeco Instruments, Inc.
Westaim, Inc.

PUNK ZIEGEL NANOTECHNOLOGY INDEX

Accelrys Software, Inc.
Altair Nanotechnologies
BioSante Pharmaceuticals
FEI Co.
Flamel Technologies
Harris & Harris Group, Inc.
JMAR Technologies, Inc.
MFIC Corp.

Nanogen, Inc.
Nanophase Technologies
Nano-Proprietary, Inc.
NVE Corp.
SkyePharma PLC
Symyx
Veeco Instruments

As has been consistently reiterated throughout this book, there is a great difference between some of the companies claiming to be nanotechnology companies and those who actually are. There is even a bigger difference between those with promising nanotechnology-related intellectual property and product "concepts" and those who are actually producing real products.

To that end, there are three distinct problems with each index of which the individual investor should beware. First, neither attempts to reflect the work that major corporations are doing in the field of nanotechnology. Their argument is that because nanotechnology makes up such a small portion of these companies overall business that they can't yet be fairly characterized as nanotechnology companies. The argument has some merit, but as Chapter 5 demonstrated, it is tenuous at best. Many of these companies are investing heavily in the field, are partnering with some of the most exciting nanotech start-ups, and/or already have real nanotechnology products on the market. The percentage of revenues attributed to nanotechnology is only going to grow in the years ahead as their nanotechnology research and development begins to pay dividends.

The second problem is that both indexes are U.S.–centric. This leaves out legitimate foreign companies such as BASF, Degussa, NEC, Obducat, pSivida, and Starpharma, which are just as involved in nanotechnology—if not more so—than many of the companies listed by either Merrill Lynch or Punk Ziegel.

Finally, neither company attempts to discern the relative quality of the stocks they list. For instance, both indexes list Altair as a nanotechnology company—which it is—but neither makes a point to note that its price-to-earnings ratios are nonexistent and its book-to-price ratio places it on the far outer edges of what constitutes a sensible investment.

Still, the indexes do serve their stated purpose of providing investors with a broad sense of nanotechnology-related activity on the market, and I encourage investors to track both indexes, if for no other reason than to see what companies are added or removed. Merrill Lynch's index is updated on an as-needed basis, while Punk Ziegel updates its index twice yearly on March 15 and August 15. Both indexes use December 31, 2003 as their base date.

What About a Nanotech Mutual Fund?

Many readers are undoubtedly curious to know if a nanotechnology mutual fund exists. The short answer is yes. Unfortunately, the largest two are limited to residents of Germany and Luxembourg, which is where the Activest Lux Nanotech and the Lux DAC Nanotech Fund are based. In March 2005, First Trust Portfolios began trading a limited nanotech mutual under the symbol FTNATX, but little information is known about its portfolio. (Since publication of this book, a new nanotechnology mutual fund has become available in the United States. Please see the Web site *www.nanonovus.com* for further information.)

Because the Activest Lux Nanotech fund represents the closest model to how I believe investors should approach investing in nanotechnology, the ten largest companies in its fund, along with the percentages, are listed as follows:

JEOL Td—6.4%	*FEI—3.8%*
Symyx—5.5%	*Headwaters—3.5%*
Input/Output—5.5%	*American Pharmaceutical Partners—3.3%*
Varian—4.9%	*Caliper Life Sciences—3.2%*
Veeco Instruments—4.8%	*Cash—16%*
Molecular Devices—3.9%	

Note: The fund has equity stakes in other companies, but the complete list was not available. Therefore, the total falls well short of 100%.

The Activest Lux Nanotech fund is described as a global equity fund for the future-oriented investor and was established on November 4, 2002. The fund manager, Thiemo Lang, has taken a fairly conservative approach by investing heavily in a number of the equipment suppliers (JEOL, Veeco, FEI, etc.) and has only a small portion in the life sciences and pharmaceutical sector (American Pharmaceutical and Caliper) and even less in energy (Headwaters). The fund is also heavily invested in the U.S. stock market.

The Lux DAC Nanotech Fund breaks its fund down into three separate fields: Nanotech Enablers, Nanotech Leading Researchers, and Nanotech Pure Plays. It has a 4 percent equity stake in each of the twenty-five companies.

NANOTECH ENABLERS (32%)

Applied Films Corporation (AFCO)	*Symyx Technologies, Inc.*
FEI company	*Varian, Inc.*
NanoMetrics	*Veeco Instruments*
NVE Corporation	*Zygo Corporation*

NANOTECH LEADING RESEARCHERS

Dow Chemical	*IBM*
DuPont	*Intel Corp.*
ExxonMobil	*Medtronic*
General Electric	*Motorola*
Hewlett-Packard	

NANOTECH PURE PLAYS

Elan Corporation	*Nanogen, Inc.*
Flamel Technologies	*Nanophase Technologies*
Harris & Harris	*Pharmacopeia, Inc.*
Headwaters Incorporated	*SkyePharma*

The fund also has the right approach and appears to have an even more balanced approach than Activest Lux Nanotech, although the fact that it invests the same amount (4 percent) in every company suggests little analysis has been applied to the relative merits of its companies vis-à-vis the others. Furthermore, its classification of certain companies as nanotechnology players (e.g., Nano-Metrics, Nanogen, and Pharmacopeia) leaves a bit to be desired.

Since both of these funds are not available to the U.S. investor, the only option at the present time is to create their own fund. Hopefully, this book will have provided the basis for doing so, but before investing in any individual stock, individual investors are encouraged to consult a few more sources because the information will be slightly out of date by the time of publication.

The first is *www.smalltimes.com*. *Small Times* is a daily news source dedicated to covering developments in nanotechnology (and MEMS). It was originally funded by Ardesta, a venture capital firm that specializes broadly in "small technology," but the media division branched off on its own in 2005. It also has an excellent (and free) monthly magazine. Investors are encouraged to use its Web site search function to track recent developments.

The second resource is *www.nanotech-now.com*. Nanotechnology Now is another wonderful free daily Web site that scans thousands of periodicals and journals and posts links to virtually every article relating to nanotechnology. It is an invaluable source for those investors who have the time to do their own due diligence.

The third free source is compliments of Cientifica, one of the larger suppliers of nanotechnology information. The company offers a weekly news report, *TNTWeekly*, which gives a quick synopsis of the week's most relevant nanotechnology-related activity and does an outstanding job at separating out the hype from any number of specious press releases. The main author, Tim Harper, also has a blog that is worth following. Both can be accessed at *www.cientifica.com*. Cientifica also publishes *The Nanotechnology Opportunity Report*. It comes with a steep price tag (2,995 euros), but it is comprehensive. Neither *Small Times, Nanotech-Now,* nor *TNTWeekly* offer analysis on individual nanotechnology stocks.

The *Forbes Wolfe Nanotech Report* does offer analysis on nanotechnology stocks, and it is excellent resource and a worthwhile investment for investors who can afford the $295 annual subscription. It is a monthly report dedicated to tracking nanotechnology stocks and maintains a running list of those companies it recommends. It also issues alerts on nanotechnology-related activity on an as-needed basis.

Interested parties may also wish to consider an annual subscription to *The NanoBio Report.* It is a weekly electronic newsletter with a more exclusive focus on nanobiotechnology and as a strong bias toward the health care arena—specifically the pharmaceutical, diagnostics, medical device, and cancer treatment industries. It, however, also comes with a steep price—$1,295.

Investors are also encouraged to visit *NanoNovus.com,* the daily blog associated with this book, which tracks nanotechnology-related developments that affect the investment and business marketplace.

Lastly, I encourage interested investors to closely monitor the activities of three venture capital firms: Lux Capital, Draper Fisher Jurvetson, and Harris & Harris. Short profiles are listed as follows. They are worth tracking because they have their finger on the pulse of the most exciting private nanotech start-ups.

► Lux Capital
245 Park Avenue, 24th Floor, New York, NY 10167
212-792-4188

Lux Capital is well versed in the nanotechnology field. One of the firm's managing partners is Josh Wolfe, editor of the monthly *Forbes Wolfe Nanotech Report* and a cofounder of the NanoBusiness Alliance. Wolfe and his team have a deep network of relationships all across the private sector, academia, and government. To date, Lux Capital has invested in three nanotechnology start-ups: Cambrios, Molecular Imprints, and Nanosys. Lux Capital is also affiliated with Lux Research, which is a research and advisory firm focusing on the business and economic impact of nanotechnology and related emerging technologies, and also publishes the more comprehensive *Nanotech Report.*

► **Draper Fisher Jurvetson**
2882 Sand Hill Road, Menlo Park, CA 94025
650-233-9000

One of the firm's managing partners, Steve Jurvetson, is referred to as "Mr. Nanotech" and is leading DJF's aggressive nanotechnology investment strategy. The quality of DFJ's research team ensures the company has done its due diligence, and its contacts within the industry help its portfolio companies make the necessary marketplace connections to increase the odds of achieving success. To date, the company has invested $80 million in the following nanotechnology companies: Arryx, Coatue, D-Wave Systems, Imago, Konarka, Molecular Imprints, NanoCoolers, NanoString, Nantero, NeoPhotonics, Intematix, and ZettaCore.

Finally, although it is publicly traded and profiled on pages 175–76, investors are encouraged to pay close attention to Harris & Harris. As was previously mentioned, the company's stock offers the only way for the individual investor at the present time to take an equity position in the aforementioned nanotechnology start-ups.

Building Your Own Fund

Every investor has a different set of objectives for investing, and each investor has his or her own tolerance for risk. The purpose of this section is not to design a one-size-fits-all nanotechnology mutual fund (although a model portfolio is offered). Rather, it is to provide the average investor with a prudent method for approaching investing in this field.

To begin, almost every financial advisor urges their clients to diversify their portfolios with a mix of bonds, stocks, and some cash. More cautious investors or those who have a shorter investment horizon are encouraged to carry a heavier mix of bonds. Those who are more aggressive or who have a longer time before retirement typically invest more in equities. It is up to the individual to determine how much, if any, of their overall portfolio they wish to invest in bonds.

The next consideration is to diversify one's stock portfolio across a variety of industrial sectors. The logic is simple. Due to the cyclical nature of many sectors, it is unwise to invest too much in any one sector on the chance that it experiences a downswing and drags down your overall portfolio performance with it.

The same logic applies to finding an appropriate balance between U.S. and foreign stocks. Political events, fluctuations in currency valuations, and a host of other economic factors can hit any one market particularly hard, and having a balance helps hedge against such factors.

Investors concerned about investing too much of their portfolios in nanotechnology stocks need not be overly concerned. While I don't advise putting all of your money into nanotechnology-related stocks, the fact is that because nanotechnology affects virtually every industry, with just little bit of attention, it is easy to maintain a healthy balance of diversity. Furthermore, by investing in some of the larger companies—such as those mentioned in Chapter 5, investors will gain some additional exposure to a diversity of markets.

As for investing in foreign stocks, this makes sense for the aforementioned reason. But it also makes sense because some of the most promising nanotechnology companies are located in Australia, France, Germany, Japan, and the United Kingdom.

The next consideration is to determine an appropriate mix between growth stocks and aggressive stocks. Growth stocks tend to be in more mature industries such as equipment, materials/chemicals, and energy. More aggressive stocks can be found in the life science/health care and information technology sectors. A further distinction can be made between large companies and small to medium-sized business. As a general rule, large cap stocks (over $1 billion in annual revenues) typically demonstrate less volatility, whereas small and mid-cap stocks have more.

Lastly, most financial advisors encourage keeping a certain amount of cash on hand in order to meet any large unexpected expenses and to hedge against unforeseen life events—such as sudden unemployment. Again, it is a prudent strategy, but for our purposes, I am going to suggest investors keep a small amount of cash on hand in the event a promising nanotechnology company goes public.

With all of these factors in mind, I therefore encourage nanotechnology investors to determine their own mix among these various categories: foreign and U.S. stocks, large and small cap stocks, and the equipment suppliers, materials, and chemicals, life sciences/health care, energy, and information technology sectors. Following are the suggested categories with a list of companies I believe will perform well in the years ahead.

Foreign Companies: *ASM International, BASF, Degussa, Flamel, JEOL, NEC, Obducat, pSivida, SkyePharma, Starpharma*
Equipment Companies: *Accelrys, FEI, JEOL, KLA-Tencor, MTS Systems, Obducat, Symyx, Varian, Veeco*

Material/Chemical Companies: *BASF, Cabot, Degussa, Dow Chemical, 3M, Oxonica*

Life Sciences/Health Care: *Affymetrix, American Pharmaceutical Products, Caliper Technologies, CTI Molecular, Flamel, Immunicon, pSivida, SkyePharma, Starpharma*

Energy Companies: *ChevronTexaco, Engelhard, Headwaters*

Information Technology: *Hewlett-Packard, IBM, Intel, NVEC, NEC*

The following is a suggested portfolio for a typical forty-year-old investor wishing to pursue a moderately aggressive investment strategy in nanotechnology.

	LARGE CAP 28%	MID CAP 33%	SMALL CAP 29%
Equipment 15%	JEOL 5%	FEI 5%	Accelrys 5%
Materials/Chemicals 15%	Dow 5%	Symyx 5%	Oxonica 5%
Life Sciences/Health Care 25%	Affymetrix 5%	Flamel 10%	Starpharma 10%
Energy 10%	ChevronTexaco 3%	Engelhard 3%	Headwaters 4%
Information Technology 25%	HP 10%	NVEC 5%	Harris & Harris* 10%
Cash 10%			

*Harris & Harris has investments across a number of industry sectors and was placed in this category because it reflects a majority of its investments.

As for the remaining 10 percent, I recommend keeping it in cash in the event any of the following companies goes public in the near future: Arryx, BioForce NanoScience, Cap-XX, Catalytic Solutions, Dendritic Nanotechnologies, Evident, Fluidigm, Imago, Konarka, Molecular Imprints, NanoCooler, NanoInk, Nanomix,

NanoSpectra BioSciences, Nanosphere, Nanosys, Nantero, nPoint, Quantum Dot, ZettaCore, or Zyvex.

The odds are that at least a few of them will return spectacular gains in the years ahead.

Conclusion

On December 17, 1903, Wilbur and Orville Wright first achieved flight. It was a historic event. Surprisingly, there was no mention of the event in newspapers the following day. In fact, most Americans weren't even aware of the event until years afterward.

Instead, the lead story in most U.S. newspapers that day was a story reporting that the automobile driving record from New York to Los Angeles had just been shattered. A fellow by the name of Colonel Jackson had reduced the record down to a remarkable sixty-three days!

I end with this story because just as it would have been virtually impossible for most citizens alive on December 18, 1903, to comprehend that the event that had transpired the day before in Kitty Hawk, North Carolina, would reduce the time it took to get from New York to Los Angeles by 99.6 percent, so too is it almost incomprehensible for today's citizens to understand that nanotechnology will shatter many industries by a comparable amount.

I encourage you to think of nanotechnology today as being at the same relative level of technical sophistication that the aerospace industry was at when the Wright Brothers achieved their modest twelve seconds of flight.

For those investors who take the time to study nanotechnology and then invest in its potential, the future is going to be an unbelievable ride, and your return on those investments could be out of this world.

Index

3M, 10, 110, 114–15, 244, 273
4Wave, 203

A

Accelrys, Inc., 41–42, 264, 272
Activest Lux Nanotech, 266, 267
Advance Nanotech, 156–57
Advanced Micro Devices, 27, 116–17
Affymetrix, 118–19, 273
Albany Nanotech, 27
Alivisatos, Paul, 29
Altair Nanotechnologies, Inc., 253–54, 264, 265
Amazon.com, 11
Amcol International Corp., 6, 264
American Pharmaceutical Partners, 154, 158–59, 266, 267, 273
AMR Technologies, 76
Angstrom Medica, 202, 244
Apollo Diamond, Inc., 2–5, 204, 242
Applied BioSystems, 38
Applied Films Corporation (AFCO), 267
Applied Nanotech, Inc./Nano-Proprietary, 255–56, 264
Argonide Corporation, 205
Arrowhead Research Corporation, 160–61
Arryx, Inc., 62, 270, 273
Asahi Glass, 5
ASM International, 40, 272
Aspen Aerogels, 81
Assessing technology, 30–31, 34–35
August Technologies, 20
Avantium Technologies BV, 63

B

Barker, Joel, 199
BASF, 10, 75, 108, 120–21, 265, 272, 273
Bell Labs, 199
Bezos, Jeff, 11
Bio Nano Research Institute, 119
BioDelivery Sciences International, 162
BioForce Nanoscience, Inc., 206, 273
Biophan Technologies, Inc., 163–64
BioSante Pharmaceuticals, 165, 264
Bishop, David, 199
Bock, Larry, 12
Boeing, 110
Bristol Myers Squibb, 154
Bush, George W., 6
Buyer beware, 34–35

C

Cabot Corporation, 122–23, 264, 273
Caliper Life Sciences, Inc., 166, 266, 267, 273
Cambrios Technologies, 207, 242, 270
Cap-XX, Inc., 82, 273
Capitalism, 244
Carbon Nanotechnologies, 29, 75, 83–84
Carbon Nanotechnology Research Institute, 119
Cash, 266, 273
Catalytic Solutions, Inc., 75, 85, 273
Celera, 38

About the Author

JACK ULDRICH is the author of *The Next Big Thing Is Really Small: How Nanotechnology Will Change the Future of Your Business* (Random House, 2003) and the editor of *NanoNovus.com*. He is also the president of the NanoVeritas Group, an independent nanotechnology consultancy group, a world-renowned public speaker, and author of scores of other nanotechnology-related articles that have appeared in *The Wall Street Journal, The Motley Fool, Small Times, The Futurist, Future Quarterly Research, TechCentral Station, Executive Excellence,* and *Leader-to-Leader.* He can be reached at *jack@nanoveritas.com.*